Essential Anatomy

To our twins

Andrew and Nicholas

A guide to important principles

Essential Anatomy

SECOND EDITION

Professor J. Joseph

MD DSc FRCOG
Professor of Anatomy
University of London

MTPPRESS LIMITED·LANCASTER·ENGLAND
International Medical Publishers

Published by
MTP PRESS LIMITED
Falcon House
Lancaster, England

Copyright © 1979 J. Joseph

First edition 1971
Second edition 1979

British Library Cataloguing in Publication Data

Joseph, Jack, b. 1913
    Essential anatomy. – 2nd ed.
    1. Anatomy, Human
    I. Title
    611   QM23.2

    ISBN 0–85200–239–4

Printed and bound
in Great Britain by
REDWOOD BURN LIMITED
Trowbridge & Esher

Contents

THIS SERIES REPRESENTS A NEW APPROACH to medical education. Each title has been written by a leading expert who is in close touch with the education in his particular field.

These books do not cover any particular examination syllabus but each one contains more than enough information to enable the student to pass his or her examinations in that subject. The aim is rather to provide the understanding which will enable each person to get the most out of and put the most into his or her profession. Throughout we have tried to present medical science in a clear, concise and logical way. All the authors have endeavoured to ensure that students will truly understand the various concepts instead of having to memorize a mass of ill-digested facts. The message of this new series is that medicine is now moving away from the poorly understood dogmatism of not so very long ago. Many aspects of bodily function in health and disease can now be clearly and logically appreciated: what is required of the good nurse or paramedical worker is a thoughtful understanding and not a parrot-like memory.

Each volume is designed to be read in its own right. However, four titles: *Physics, Chemistry and Biology*; *Anatomy*; *Biochemistry, Endocrinology and Nutrition* and *Physiology* provide the foundations on which all the other books are based. The student who has read these four will get much more out of the other books which relate to clinical matters.

Preface

Anatomy to most people is a subject which suggests the cutting up of dead bodies (the word literally means *cutting up*). In addition it is generally known that Vesalius published a book in 1543 in which much of the human body was described in detail and more or less accurately. A subject which is dead and ancient frequently has little appeal especially if it appears to involve learning a large amount of factual information. For many years anatomy has had to struggle with these disadvantages and at times one has had the impression that there is almost a conspiracy on the part of everyone to suggest that anatomy is unnecessary. There is no doubt, however that a knowledge of the structures of the body, for that is what anatomy is, whether it is what can be seen with the naked eye or with different kinds of microscope, is an essential preliminary and corollary to the understanding of the functions of the body. It was no historical accident that Vesalius, the anatomist, preceded Harvey, the physiologist. No apology need be made for trying to present the basic facts of anatomy to anyone interested in the human body and to members of any profession which will have to cope with the physical and mental problems of children, men and women in health and in sickness. It is not intended that the reader should know everything contained in this book. It is hoped, however, that with the help and guidance of teachers, a comprehension of the structure and function of the human body will be acquired more easily if more information is available than the bare minimum. On the whole, the criterion used for including any information is whether it is likely to help the understanding of how the human body works, and also to stimulate interest for further study.

My thanks are due to Miss Mary Waldron and Mrs. Carol Dawbarn for the drawings and Mr. D. G. T. Bloomer of MTP Press Limited for his cooperation in producing the book.

ix

Introduction

The scheme of this book will not differ very much from that of many other textbooks of anatomy with perhaps one important difference. There will be a special emphasis on the functions of what is described. Inevitably one must begin with the *cell*, which is regarded as the basic unit of any living, complex organism. Cells form *tissues* which are aggregated to form *organs*. The cells of different organs vary in their structure and organization, and associated with this variation are the different functions they perform. Complexity of the organism has also resulted in different parts having different functions and where one group of structures has similar functions one refers to a *system* of the body, for example, the heart and blood vessels constitute the *cardiovascular* or *circulatory system*.

Terminology in human anatomy

Certain terms are used for descriptive purposes in human anatomy. The *anatomical position* is one in which the body is upright with the palms of the hands facing forwards and the feet pointing forwards. *Medial* means nearer to and *lateral* means further from the midline, that is the thumb and little toe in the anatomical position are lateral and the little finger and big toe are medial. The front is *anterior* (*ventral*) and the back is *posterior* (*dorsal*). There are three planes, *sagittal*—a vertical plane anteroposteriorly, *coronal* (*frontal*)—a vertical plane from side to side, and *transverse*—a horizontal plane (Fig.1). *Superior* (*cranial*) means nearer the head and *inferior* (*caudal*) nearer the feet. In the limbs *proximal* and *distal* are often used as synonyms for superior and inferior. *Superficial* means nearer the surface of the body as opposed to *deep* meaning further from the surface.

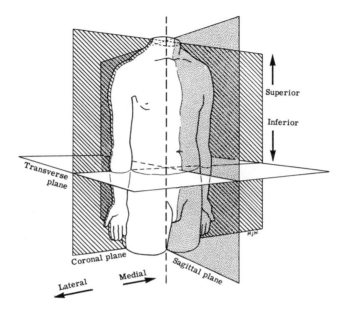

Fig.1. Terminology in human anatomy.

1

Cells

It is not easy to distinguish between living and non-living matter. If one has to make a distinction perhaps living matter may be said to have the capacity to use energy, excrete, reproduce itself and pass on to what it has reproduced its own characteristics (heredity). If this definition is accepted the lowest form of living matter is probably a *bacterium*. A virus can be regarded as an inter-mediate form between living and non-living.

The cell

The unit of living matter is called a *cell* (Fig.2). For practical purposes all cells consist of *cytoplasm* containing a *nucleus*. The nucleus consists largely of *water*, and *nucleic acid* and *proteins* in combination. The cytoplasm consists largely of water containing proteins which are found in many of the structures referred to below, for example, mitochondria, ribosomes. The cell is surrounded by a *cell membrane* and the nucleus by a *nuclear membrane*, both containing protein and lipid. Viruses consist almost entirely of nuclear material containing *nucleoproteins* and are able to use the proteins manufactured for them by the cell. For this purpose a virus has to use the *enzymes* of the cell (enzymes are similar to *catalysts*—they hasten a chemical change without taking part in the chemical change itself). Viruses therefore live inside other cells. They may destroy these cells or they may live in harmony with them. The former occurs in many diseases; for example, poliomyelitis (infantile paralysis) is due to a virus infection of certain nerve cells in the spinal cord and possibly the brain. Bacteria, bigger than viruses, have their own enzyme systems and can therefore exist independently of other cells.

The nucleus of a cell contains *deoxyribonucleic acid* (DNA) which

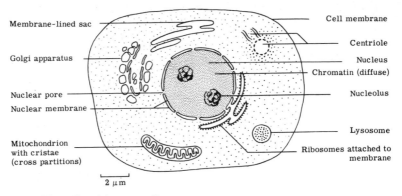

Membrane-lined sac — Cell membrane — Centriole — Golgi apparatus — Nucleus — Chromatin (diffuse) — Nuclear pore — Nucleolus — Nuclear membrane — Lysosome — Mitochondrion with cristae (cross partitions) — Ribosomes attached to membrane

2 μm

Fig.2. The cell and its organelles.

is responsible for the production of *ribonucleic acid* (RNA). Within the nucleus is the *nucleolus* (there may be more than one) which is rich in RNA. These nucleoli are the centres of ribosome synthesis. The DNA of the nucleus is localized in the *chromatin* of the nucleus. This chromatin is not easily seen unless the cell is preparing to divide, when the *chromosomes* which contain the hereditary charac-teristics of the cell become visible. The DNA determines the pattern of the proteins to be formed and the RNA is actually responsible for their formation. It passes from the nucleus through pores in the nuclear membrane into the cytoplasm of the cell. Here the nuclear RNA becomes attached to the cytoplasmic *ribosomes* where specific proteins are formed from more elementary compounds called *amino acids*, the source of which is digested food. These enzymes are found in relation to the ribosomes. Enzymes are also found in the *mito-chondria* of the cytoplasm, and function in the building up of energy-rich compounds. These energy-rich compounds are used in all activities of the cell. During this process oxygen is usually required and waste is produced in the form of water and carbon dioxide. The role of proteins in the structure and function of the cell can now be appreciated. *Lysosomes* contain enzymes which break down substances and would have this effect on the cell itself were they not contained in a membrane.

There are other structures in the cytoplasm such as the *centriole* which is involved in the division of the cell and the *Golgi apparatus* which is important in the formation of a cell's secretion. The cell membrane is so constructed that it can determine what passes into and out of the cell. This is called a *selective membrane*.

Both plants and animals consist of cells but the main distinguishing feature is that plants possess a green pigment, *chlorophyll*, which is used in a process called *photosynthesis*. By this process, which utilizes solar energy, the plant makes its own food from water and carbon dioxide. There are some unicellular organisms which use both photosynthesis and digestion of complex substances as means of obtaining food.

Certain terms are used in order to measure very small structures such as viruses, bacteria and cells. A μm (μ is the Greek letter 'm') is 1/1,000 of a millimetre and an nm is 1/1,000,000 of a millimetre, or 1/1,000 of a μm. A virus is about 50 nm in diameter, a bacterium is about 1 μm in diameter and an average cell is about 15 μm in diameter. Obviously there are large and small cells, etc. Most viruses and some of the structures in the cell referred to above can be seen only with an electron microscope which can magnify 1,000 to 200,000 times. Bacteria and cells can be seen with a light microscope which can magnify up to 1,000 times.

Single-cell organisms

It is useful to know something about the properties of a living cell. This may be best studied in a single-cell organism such as the *amoeba*.

   a. Growth: an amoeba can increase in size up to a point. This is due to its being able to undergo

   b. Metabolism: this process involves (1) the ingestion, (2) the digestion, (3) the assimilation of food. Metabolism occurs not only in growth but in almost every function of the cell, including repair. The process of metabolism involves the use of energy, the participation of enzymes and the elimination of waste products. In other words, if an amoeba were unable to ingest the right food, break it down by utilizing a source of energy with the help of enzymes, build up the broken-down products to form the substances it requires and get rid of any waste products produced in these processes, then the amoeba would die. The utilization of oxygen and production of carbon dioxide is usually called *respiration*.

   c. Reproduction: the amoeba reproduces by dividing into two daughter cells. In this process the nucleus divides into two in such a way that the daughter cells are the same as the mother cell except for size.

   d. Motility: amoebae move by pushing out some of their cytoplasm and retracting another part, that is, by the formation of
   pseudopodia.
   e. Reacting to the environment: this includes the movements
   involved in engulfing a particle of food as well as reactions such
   as movement away from an obstacle.

Many of the functions of the cell are directed towards maintaining
its equilibrium in a variable environment, that is, the cell exhibits
homeostasis but there is obviously a limit to the changes to which
the cell can react without damage or destruction.

   Unicellular organisms are classified as *protozoa*. It is interesting
to consider another type of unicellular organism in which different
parts of the same cell have become specialized for certain functions,
the *paramecium*. Unlike the amoeba, this organism has a permanent
shape with a front and back end. Food is ingested at only one place
and not anywhere on the surface. Movement is by means of *cilia*—
hairlike structures projecting from the surface and beating rhythmically. Reproduction is preceded by an exchange of nuclear
material between two paramecia followed by division. Respiration
and excretion, however, takes place over the whole surface.

Multiple-cell organisms

In multicellular organisms *specialization* takes place so that groups
of cells have a different structure and function. Cells are specialized
for reproduction, movement, digestion, excretion and receiving
information (both from the environment and the body itself). In
addition, a single function may require a complex arrangement; for
example, the conveying of oxygen to the tissues requires a pump and
an involved system of tubes. There are many ways of subdividing
the millions and millions of cells which constitute any animal.
They may be subdivided into *tissues* because they have a common
origin or function or position. Cells also constitute *organs* which are
usually mixtures or different tissues or they may form *systems* because
the various parts serve a fairly common function.

2

Tissues

Classification of tissues presents a problem because it is often difficult to find a basis which rigidly confines certain structures to one group. Traditionally tissues are said to be:

    a. Epithelial when on a surface either outside or inside the body,

    b. Connective when they form packing and supporting tissue,

    c. Muscular if they are contractile,

    d. Nervous if they possess conductivity.

This classification appears to depend on site or function. It leaves out fluid such as *blood* and *lymph*. It disregards embryological origin (for example, epithelia are ectodermal, mesodermal or endodermal in origin, all connective tissue is mesodermal and almost all nervous tissue is ectodermal). It also appears to exclude certain specialized tissues, for example, glandular and lymphoid tissue. The former is derived from epithelial tissue as a rule and the latter may be included among connective tissue.

Epithelial tissue

This tissue covers a surface which may be external (as in the skin) or internal (in the mouth or alimentary tract) or lining a vessel (blood or lymphatic). Epithelial tissue is said to be *simple* or *compound* (Table 1).

SIMPLE. This type consists of a single layer of cells whose shape gives the epithelium its name.

    a. Squamous (*pavement*) (Fig.3a) has flattened cells and is found lining blood vessels and the chamber of the heart (it is usually called *endothelium* in these structures).

5

Table I

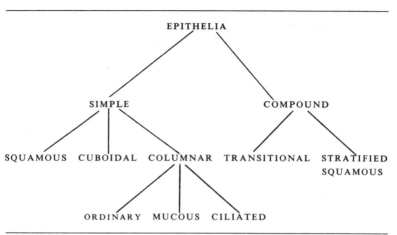

b. *Cuboidal* (Fig.3b) has roughly cube-shaped cells and is found
    lining the ducts of some glands and the acini of the thyroid
    gland.

c. *Columnar* (Fig.3c) consists of taller cells and is found lining
    the stomach and intestines. This type frequently includes cells
    (*goblet cells*) which produce *mucus* (Fig.3d) and the epithelium
    is then called *mucous columnar*. If the cells have hair-like
    processes (cilia) projecting from their free surface it is called
    ciliated columnar epithelium (Fig.3e).

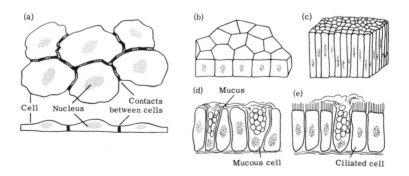

Fig.3. Simple epithelia: (a) Squamous, (b) Cuboidal, (c) Ordinary columnar,
(d) Mucous columnar, (e) Ciliated mucous columnar.

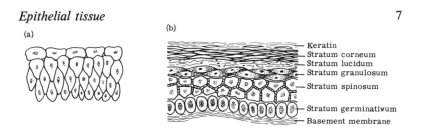

Fig.4. Compound (stratified) epithelia: (a) Transitional, (b) Keratinized stratified squamous.

COMPOUND. In this type of epithelium there are several layers of cells. There are two varieties of compound (*stratified*) epithelium.

   a. Transitional (Fig.4a), found only in the urinary passages, mainly the ureter and bladder, consists of about three to five layers of cells which are frequently pear-shaped. When the bladder is distended the epithelial lining is very much thinner than when the bladder is small. This change in thickness is one of the characteristics of transitional epithelium.

   b. Stratified squamous is called this somewhat confusing name because there are several layers, the most superficial of which are flattened (Fig.4b) The deepest layer of this epithelium is the source of the cells of all the other layers and it consists of large columnar cells. The middle layers are more polyhedral and the most superficial layer is flattened. This type of epithelium is found in the mouth and oesophagus. This epithelium is also found in the superficial part of the skin and in this position the most superficial layer consists of a layer of flattened dead cells containing *keratin*. Hence the terms *keratinized* and *non-keratinized* stratified squamous epithelium. Keratin is a hard, horny-like substance and has a protective function. The formation of the keratin can be traced through the successive layer of cells as changes in the granules in the cytoplasm of the cells forming these layers.

The keratinized stratified squamous epithelium of the skin is called the *epidermis*. The deeper layer of the skin is called the *dermis* and consists of connective tissue. *Nails, hairs* with their associated *sebaceous glands* and *sweat glands* are all modifications of the cells of the epidermis (Fig.5). The hairs and sweat glands are downgrowths of epidermal cells into the underlying dermis and the sebaceous glands develop from the cells which form the hair. Stratified squamous epithelium is often called the *wear and tear* epithelium because it is found where the surface cells are easily removed by injury or rubbing.

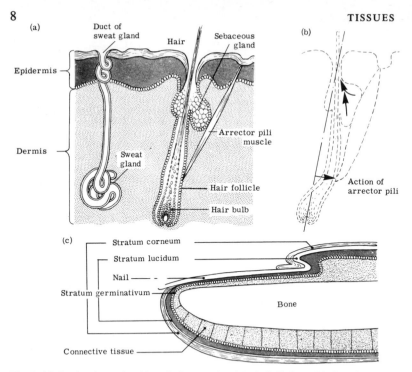

Fig.5. (a) Derivatives of epidermis (sweat gland, hair follicle and hair, sebaceous gland), (b) Action of arrector pili muscle (straightening of hair, pulling down of skin as in goose pimples, possibly squeeze sebaceous gland), (c) Longitudinal section through terminal part of finger to show relation of nail to epidermis.

Glandular tissue

Because these structures usually develop from and are often related to epithelial tissues they are dealt with at this stage. Glands which open on to the surface either directly or through a duct are called *exocrine* as opposed to the *ductless* (*endocrine*) *glands* whose secretion passes directly into the blood stream. The goblet cells of a mucous columnar epithelium may be regarded as a single-celled gland. If, as in the stomach, the cells grow down into the underlying tissue a *simple tubular gland* (Fig.6a) is formed. In another type of gland the down-growing tube branches at its end and subsequently only the branches become secretory. This is called an *alveolar gland* (Fig.6b). In some glands secretory cells are arranged round the duct in a spherical or tubular manner. Sometimes the ducts branch before the secretory tissue is formed and if the secretory tissue is racemose in its arrangement the gland is called a *compound racemose* or *alveolar gland* (racemose = like a bunch of grapes) (Fig.6c).

Glands are normally named after the type of secretion they pro-

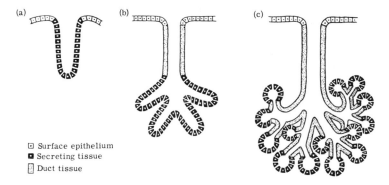

□ Surface epithelium
■ Secreting tissue
▣ Duct tissue

Fig.6. (a) Simple tubular gland, (b) Alveolar gland, (c) Compound alveolar gland.

duce, for example, *salivary glands, sweat glands, sebaceous glands.* The cells of glandular tissue may produce their secretion without much change in the basic structure of the cell (*merocrine glands*). This is commonly found throughout the body. On the other hand the luminal part of the cell may break off altogether with the accumulated secretion (*apocrine glands*). This is found in the sweat glands of the arm-pit. Sometimes the whole cell disintegrates (*holocrine glands*). This is found in the sebaceous glands of the skin.

During the production of secretion the mitochondria of the cell are very active and the nucleus undergoes changes both in position (it moves towards the base of the cell and away from the luminal edge) and structure. The secretory granules or globules accumulate near the luminal side of the cell before being released in one of the ways described above.

Connective tissue

Basically connective tissue consists of *cells* embedded in a *ground substance* containing *fibres* (Fig.7a). The state of the ground substance and its composition, and the type and quantity of the fibres determine which type of connective tissue one is dealing with. The cells are all derived from one basic type of cell called a *mesenchyme cell* Subsequently many types of cell are found and these are named according to their main function. A variety of cells and fibres, however, may be found in certain types of connective tissue which will be classified later on.

CELLS. A *fibrocyte* (Fig.7a) is the basic cell found in *fibrous tissue.*

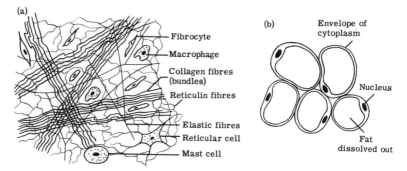

Fig.7. (a) Generalized connective tissue, (b) Fat cells.

It is a somewhat elongated cell with pointed ends and contains a cylindrical or oval nucleus. When actively producing fibres the cell is called a *fibroblast*. It is then more irregular in outline and has long tapering processes. (A cell name ending in *-blast* usually implies an active phase of the cell as compared with a name ending in *-cyte* which implies an adult or the final resting stage of a cell's development.) There are various cells in connective tissue which are *phagocytic*, that is, they are capable of ingesting particles. These are called *macrophages* or *histiocytes* and may be *fixed* or *wandering*. *Reticular cells* are potential phagocytes in that they can become free and phagocytic. These cells also produce fine fibres called *reticulin*. The cells themselves have long processes which may join up and form a network together with the fibres they produce. Hence their name. *Fat cells* (Fig.7b) are, as their name suggests, cells which contain fat globules. Usually they contain so much fat that the cytoplasm forms a thin envelope and the nucleus is pushed to one side. *Mast cells* are larger than the others with deeply staining granules. They produce *heparin*, a substance which prevents clotting of the blood.

GROUND SUBSTANCE (MATRIX). This consists of *tissue fluid* containing additional substances which are complex proteins. The state of the tissue matrix, that is, whether it is more or less viscous, depends on the *polymerization* of the proteins. (Polymerization refers to a change in which a large number of small molecules becomes a small number of large molecules.) Tissue fluid contains water, salts, glucose, albumin, etc. in varying quantities. The constitution of the ground substance can vary a great deal.

FIBRES. Some look *white* and are also known as *collagen*. They are thin and wavy and lie in bundles. With an electron microscope these fibres are seen to have transverse light and dark bands. Collagen is formed by fibroblasts and when aggregated into bundles has considerable tensile strength. *Reticular fibres* are much thinner than collagenous fibres and branch to form fine networks. They are formed by the reticular cells. It is believed that collagen is formed from reticular fibres. *Elastic fibres* look *yellow* and are sometimes wavy and may be fairly thick. As their name suggests they are capable of being elongated without breaking. Their source is not known.

TYPES OF CONNECTIVE TISSUE
Packing.

Reticular tissue, as the name suggests, consists of a reticular network of fibres with reticular cells in its meshes. The cells and fibres lie in a matrix which is frequently more fluid than solid. This type of tissue is rarely found alone. It often forms part of the framework of organs such as the lungs and the liver and of structures such as lymph nodes.

Loose and *dense connective (fibrous) tissue* consists of a varying amount of collagen bundles containing many of the cells named above in a more or less viscid matrix. Elastic and reticulin fibres are also found but not in such large quantities. This tissue, if loose, is also known as *areolar* and then may form the packing tissue of various organs of the body. Areolar tissue often forms a layer of tissue beneath an epithelium, for example the submucosal layer of the mucous membrane of the alimentary tract. In this are found small vessels and nerves. Loose connective tissue deep to the skin is also known as *superficial fascia* and frequently contains a variable quantity of fat. If the tissue consists predominatly of fat it is called *adipose or fatty tissue.* In dense connective tissue the collagen bundles are very numerous and the matrix is very much reduced in quantity. The cellular elements and the blood vessels are fewer than in loose connective tissue. Dense connective tissue often forms sheets. It is present as a more or less continuous layer deep to the superficial fascia and is called the *deep fascia.* It may form part of a flat muscle and is then known as an *aponeurosis.*

There are structures which consist of large numbers of closely packed collagen bundles lying parallel to one another. They are particularly strong and are found at the ends of muscles attached to bone in the form of *tendons*, and at joints where two bones are held together by *ligaments*. In this type of connective tissue the other elements are greatly reduced and there are few blood vessels and nerves.

Dense connective tissue is also found as an ensheathing layer round muscles, nerve trunks and blood vessels. The surface of these ensheathing layers as well as of the deep fascia is smooth and this enables the structures to glide on one another.

It should be realized that many structures such as *serous membranes* (the lining of the wall of the body cavities), *synovial membranes* (found in joints lining the capsular ligament holding the two bones together) and *bursae* (fibrous tissue structures consisting of an outer wall of dense connective tissue, lined by synovial membrane and containing a sticky fluid) are all modified connective tissue structures.

Supporting

Cartilage and *bone* are two highly specialized types of connective tissue. It is of some interest that in certain circumstances cartilage and/or bone can develop in any type of connective tissue.

Cartilage. This tissue is said to be of three types but in all of them there are cells (*chondrocytes*), ground substance, which is solid and pliable, and fibres which may be collagen or elastic. Chondrocytes develop from mesenchyme cells and are spherical or oval in shape. Characteristically they are found in pairs or in groups of four or in rows and the adjacent surfaces of the cells are flattened. The *lacunae* are the spaces in the matrix in which the cells lie. The matrix contains *chondroitin sulphate*, a mucopolysaccharide (Fig.8).

In *hyaline cartilage* the lacunae containing the cells are scattered throughout the matrix which, although containing very fine fibres, stains uniformly. This type of cartilage is usually surrounded by a dense layer of fibrous tissue called the *perichondrium*, the outer layers of which are mainly fibrous. The inner are more cellular and these cells may be the precursors of chondrocytes. These inner cells are capable of producing new cartilage. This type of cartilage is seen in the *costal cartilages*. *Articular cartilage* is also hyaline but is structurally more complex and has no perichondrium.

Fibrocartilage which consists largely of dense white fibrous tissue with scattered chondrocytes is seen in the discs between the vertebrae. *Elastic cartilage* contains a large number of elastic fibres and is found in the *pinna* of the ear and the *epiglottis* of the larynx. This type is more readily deformable than hyaline which in turn is more deformable than fibrocartilage. Cartilage is a tissue which has been compared to a motor tyre which deforms easily due to the rubber and returns to its original shape, and is very strong because of the metal wires incorporated in the rubber. The ground substance of the cartilage corresponds with the rubber and the collagen fibres with the metal wires.

Cartilage is also found as a precursor to bone. It is avascular and frequently areas of calcification are found in it in later years (over 40).

Bone. This tissue is characterized by the rigidity of its matrix. This is due to the deposition of calcium and magnesium salts of phosphate and carbonate in the matrix in which fibres lie. It is possible for the salts in the matrix to be dissolved out so that a soft pliable structure remains. This consists largely of the collagen fibres which are part of the bone structure. On the other hand the fibres can be removed so that only the matrix impregnated with salts is left. This is very brittle.

Basically the fibres and matrix are arranged in a concentric tubular pattern. The tubules consist of lamellae of fibres which run obliquely in an individual lamella and in different directions in adjacent lamellae. The lamellae are divided into:

  a. Circumferential which are parallel to the outer surface of the bone

  b. Haversian which are concentrically arranged round a small central canal forming Haversian systems (osteons) and

  c. Interstitial which lie between (a) and (b) (Fig.9).

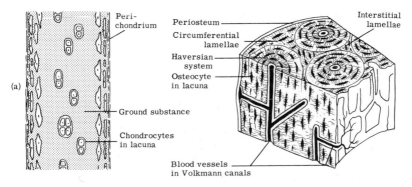

Fig.8. Hyaline cartilage.   **Fig.9.** Structure of bone.

A *Haversian system* consists of the concentric lamellae, the central canal with its contents (an artery, vein and nerve) and bone cells called *osteocytes*, between the lamellae. These cells lie in spaces (*lacunae*) between the lamellae and early on had processes which extended through small channels (*canaliculi*) in the lamellae. Haversian systems if followed longitudinally are found to branch and join one another. *Volkmann canals* pass horizontally through the Haversian systems and carry blood vessels from the periphery to the Haversian canals.

The outer surface of a bone is covered by a layer of fibrous tissue, the *periosteum*, which superficially is more fibrous and contains cells like fibrocytes, and more deeply is more cellular and contains cells which are capable of forming bone.

Most bones on section can be seen to be denser on the outside and more open on the inside. The former is called *compact* (*ivory*) bone and the latter *cancellous* (*spongy*) bone. Structurally, cancellous bone has no Haversian systems and the lamellae form trabeculae which give the appearance of a branched lattice work. In the meshes of cancellous bone is found *red bone marrow* as a rule (site of the formation of many blood cells). Many long tubular bones have a longitudinal central canal, the *medullary cavity*, which contains *yellow bone marrow* (a store-house of fat).

Bones are frequently classified according to their shapes. *Long bones* consist of an elongated shaft with an enlargement at each end. These are found typically in the arm and leg. *Short bones* are roughly cuboidal in shape and are found in the wrist and foot (*carpal* and *tarsal bones*). *Flat bones* have a considerable area in relation to their thickness. These are found typically in the vault of the skull. *Irregular bones* are of varied shape and include the vertebrae. There is another group of bones called *sesamoid* which develop in tendons of muscles or the capsular ligament of joints. The knee-cap (*patella*) is an example of this type of bone.

Ossification is the development of bone in pre-existing connective tissue. If this occurs without the preliminary formation of cartilage it is known as *intramembranous ossification*. If ossification is preceded by a cartilaginous model it is known as *cartilaginous* (*endochondral*). The former produces what are known as membrane bones and the latter cartilaginous bones. Membrane bones are found in the vault of the skull and part of the facial skeleton. The base of the skull, the vertebral column, the ribs and the limb bones are developed in cartilage.

In intramembranous ossification an area of connective tissue, consisting of collagenous fibres and fibroblasts, becomes more vascular. At the same time the matrix becomes clear and jelly-like and the fibroblasts become osteoblasts. The osteoblasts by means of an enzyme called *alkaline phosphatase* cause the deposition of calcium salts in the matrix, and the combination of the rigid matrix and the fibres within the matrix can be called bone. The osteoblasts themselves become enclosed in the bone which has been formed and become osteocytes. Removal and 'shaping' of the bone follow so that regular lamellated bone is produced. It is said that large multinucleated cells called *osteoclasts* are responsible for the removal of bone.

In cartilaginous ossification the cartilage has to be removed before ossification begins. Actually the two processes overlap to some extent. In a typical long bone ossification begins in the centre of the shaft (*diaphysis*) about the 8th week of intra-uterine life (*primary centre of ossification*) and the shaft is completely ossified by birth. The ends of the bone (*epiphyses*) begin to ossify after birth, with one or two exceptions (*secondary centres of ossification*). While growth continues the epiphysis is separated from the diaphysis by a cartilaginous plate, called the *epiphyseal cartilage*. The first change takes place in the region of the primary centre of ossification where the peripheral part of the cartilage (formerly perichondrium now periosteum) lays down bone so that the degenerating cartilage is surrounded by a sleeve of bone (Fig.10).

Changes also take place in the cartilage (Fig.10). The cartilage cells are arranged in rows parallel to the long axis of the bone. These enlarge and degenerate and the matrix of the cartilage become calcified. The calcified cartilage is invaded by blood vessels and removed so that spaces containing vascular connective tissue are formed. In these spaces bone is laid down by the connective tissue cells which can now be referred to as osteoblasts. At the ends of long bones ossification in the epiphysis occurs and is a similar process to the one described. The surface cartilage at the end of the bone, however, remains as the articular cartilage. At the epiphyseal cartilage next to the shaft the process of ossification goes on for a much longer time so that the bone elongates. With the ossification of this cartilage growth ceases.

It is important to note that growth in the length of a long bone is due to cartilaginous ossification, but growth in the width of the shaft is by membranous ossification and is due to the activity of the deepest

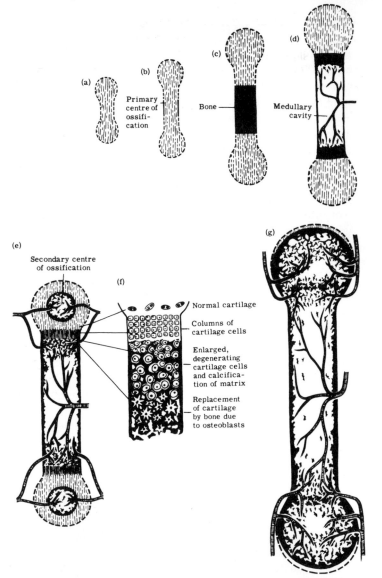

Fig.10. Cartilaginous (endochondral) ossification of a typical long bone: (a) Cartilaginous model (about 6 to 7 weeks of fetal life), (b) Primary centre of ossification round shaft (about 8 weeks of fetal life), (c) Ossification spreading in shaft, (d) Formation of medullary cavity and vascularization of shaft (before birth), (e) Vascularization of ends of bone (epiphyses) and commencing ossification of epiphyses, (secondary centres of ossification); separation of shaft from epiphyses by cartilaginous epiphyseal plate (growth plate), (f) Details of growth plate, (g) Adult bone showing disappearance of growth plates (fusion of epiphyses with shaft) at 16 to 18 years.

layers of cells of the periosteum. The growth of a bone is accompanied by the removal of bone. The simplest example of this is seen in the shaft. While bone is laid down on the outside the inside is removed so that the medullary cavity becomes wider. Similarly the ends of long bones retain their shape as growth takes place. This process is known as *modelling* (Fig.11a,b).

The final shape of a bone depends on many factors. Experimental embryology has shown that the main anatomical features of an adult bone will develop in the undifferentiated mesoderm of a limb bud which has been removed from its normal environment. Thus genetic factors play an important part. On the other hand, bones do not develop normally unless their environment provides several important substances in the right quantities at the right time. Among these are several of the vitamins (vitamins A, C, D), mineral salts (calcium and phosphorus) and hormones from the parathyroid, pituitary and thyroid glands and gonads. Mechanical factors influence the shape of a bone. Tension exerted by muscles and ligaments and the pressure of other tissues produce projections and depressions of various shapes and sizes on the adult bone. Unusual tensions and stresses can alter the whole shape of a bone. The internal structure of a bone, that is, the pattern of the lamellae, although to some extent genetically determined, is also influenced by the tensions and stresses to which the bone is subjected. This is most beautifully illustrated in a coronal section of the head, neck and upper end of the shaft of the adult femur (Fig.12).

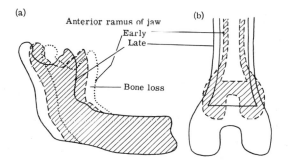

Fig.11. (a) Modelling of lower jaw to accommodate permanent teeth, (b) Modelling of lower end of femur during growth in length and width.

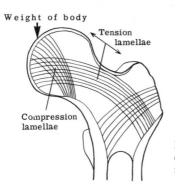

Weight of body

Tension
lamellae

Compression
lamellae

Fig.12, Tension and compression stresses
determine pattern of lamellae in head and
neck of femur.

Muscular tissue

All cells in the body in certain circumstances can contract but certain
cells, called muscle, are specialized for this purpose. These cells are
usually elongated and contract rapidly so that movement results.
They can also increase tension to resist external forces or expansion
of a cavity. Muscular tissue is classified according to whether the
muscle fibres show transverse striations or not, that is, there is
striated (striped) and *smooth (unstriped) muscle.* Striated muscle is
also called *voluntary* because it is found in those parts of the body
which are under voluntary control. It is not a good term. Similarly
smooth muscle is called *involuntary* because it is not under voluntary
control. These latter terms not only do not describe the way in which
all striated and smooth muscle functions but make the description
of *cardiac (heart) muscle* difficult because it is striated and involuntary.

STRIATED MUSCLE. A muscle fibre of this type is elongated with
somewhat tapering ends (Fig.13c). Its length is very variable (it
may be up to 30 cm long) and each fibre is about 100 μm in diameter
although the maximum length of these muscle fibres is still somewhat of a mystery. There are many nuclei in a single fibre and these
are situated peripherally. It is not known whether this type of muscle
fibre is the result of the fusion of single cells or the frequent division
without separation of one cell.

A muscle fibre usually has a connective tissue sheath (*endomysium*)
within which is the muscle membrane called the *sarcolemma*. Within
this membrane is the semifluid *sarcoplasm* in which are longitudinally arranged structures called *myofibrils*. These myofibrils have

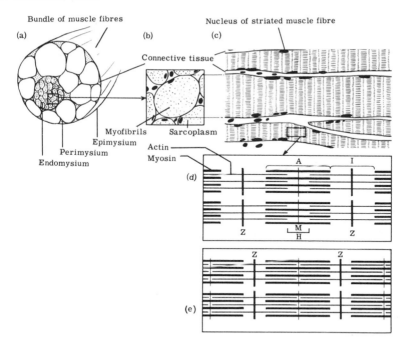

Fig.13. (a) Transverse section of whole striated muscle, (b) Transverse section of single muscle fibre, (c) Longitudinal section of single muscle fibre, (d, e), Diagrams of electron microscope appearances of uncontracted (d) and contracted (e) myofibrils.

transverse striations which are alternately dark and light. The former are called A *bands* and the latter I *bands*. Within the A bands is a light band (H *band*) and within the I band is a dark line (Z *line*). The H band and Z line are much narrower than the A and I bands. A *sarcomere*, the functioning unit of the myofibril, extends from one Z line to another Z line (Fig.13c,d). The A and I bands in an uncontracted muscle fibre are about equal in length. After contraction the A band remains the same size but the I bands are very much smaller. Electron microscopy has shown that there are longitudinal filaments in both the I and A bands. Those of the I band consist of a protein *actin* and those of the A band consist of *myosin*, also a protein. In the uncontracted muscle fibre the actin filaments are seen to lie between the myosin filaments. During contraction the actin filaments move towards each other between the myosin (Fig. 13d,e). A single muscle fibre also shows transverse striations because all the A bands of the individual myofibrils lie in line with each other.

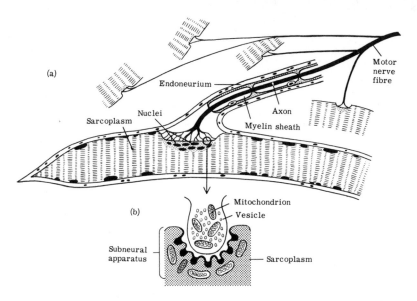

Fig.14. (a) Diagram of motor end plate, (b) Diagram of electron microscope appearance of region between nerve ending and muscle fibre.

Each muscle fibre is also supplied by at least one terminal branch of a nerve fibre. Because it is related to movement this type of nerve is called a *motor nerve*. The nerve ends at the muscle fibre in a region known as the *motor end plate* or *myoneural junction* (Fig.14a,b). This is usually seen as a depression on the muscle fibre and is characterized by an aggregation of nuclei. The terminal arborization of the nerve fibre remains separate from the sarcoplasm next to which it lies, and between the terminal filaments of the nerve fibres and the sarcoplasm is the *subneural apparatus* rich in an enzyme called *cholinesterase*.

Bundles of muscle fibres within a muscle are surrounded by a sheath of connective tissue (*perimysium*) and the whole muscle itself is surrounded by a fairly dense connective tissue sheath (*epimysium*). Within the whole muscle, in addition to the muscle fibres, there are structures called *muscle spindles* which as their name suggests are spindle-shaped (Fig.15). They are up to 1 cm long and consist of a number of small striped muscle fibres enclosed in a connective tissue sheath. In the middle of the spindle there are many nuclei (*nuclear bag* (Fig.15)) round which there is a space, containing lymph. The muscle fibres of the spindle are called *intrafusal* and the

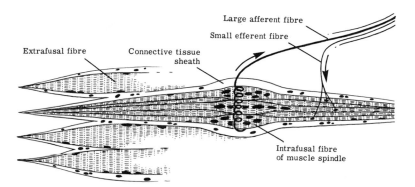

Large afferent fibre

Small efferent fibre

Extrafusal fibre

Connective tissue
sheath

Intrafusal fibre
of muscle spindle

Fig.15. Diagram of muscle spindle showing main sensory and motor innervation.

large fibres of the rest of the muscle are called *extrafusal*. Intrafusal fibres have a nerve supply similar to that of the extrafusal fibres but it is their *sensory* function, that is, the impulse they convey *from* the muscle, which is very important. Penetrating the fibrous capsule of a muscle spindle and winding round the middle of the spindle is the large peripheral process of a sensory neurone with its cell body in a posterior root ganglion and its central process going to the spinal cord in a posterior nerve root. Stretching of the whole muscle or contraction of the muscle fibres in the spindle sends an impulse along this sensory neurone. By this means information about the extent to which muscles are being stretched is conveyed to the nervous system. There are also branched sensory nerve endings lying among the muscle fibres near the attachment of the muscle to its tendon. These also convey information to the central nervous system when the muscle is stretched. These impulses are inhibitory.

SMOOTH MUSCLE. The fibres of this type of muscle are smaller (100–200 μm in length) than those of striated, and are unbranched. They taper at each end and have one central nucleus in each fibre (Fig.16a). There are no transverse striations in these fibres although there are longitudinal myofibrils in the cytoplasm of the cell. Smooth muscle often forms sheets. There are frequently reticular and collagenous fibres between bundles of smooth muscle fibres and even between individual fibres. This type of muscle is found in viscera (for example, the lungs, stomach, uterus, bladder) and is often called *visceral* muscle. In the alimentary tract it forms circular and longitudinal layers.

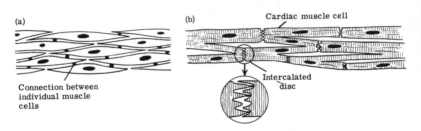

Fig.16. (a) Smooth muscle fibres, (b) Cardiac muscle fibres.

Smooth muscle is innervated by *autonomic* (*involuntary*) *nerves.* (The motor nerves supplying striated muscle are called *somatic.*) Although viscera have a sensory nerve supply, there is nothing comparable in smooth muscle with the muscle spindle of striated muscle.

CARDIAC MUSCLE. As the term implies this type of muscle is found in the heart. It is striated and involuntary. Essentially the cells of heart muscle act like a *syncytium*, a more or less continuous mass of tissue. Individual cells are about 75 μm long and 20 μm wide with a central nucleus, but they lie in series, and branch and join with each other. There are much-folded membranes between the ends of cells. These are called *intercalated discs* (Fig.16b). Each cell contains longitudinal myofibrils whose structure is very similar to those of striated muscle (A and I bands, etc.).

Some of the muscle of the heart is modified and forms what is known as the *conducting system* of the heart. This will be described with the section on the heart itself. Heart muscle has a very rich blood supply and is supplied by a large number of autonomic nerve fibres.

Nervous tissue

It has already been remarked that the amoeba responds to certain stimuli, that is, the cell possesses irritability. Another way of expressing this is to say that the cell responds to its environment. Nervous tissue in more complex animals is that special part of the body which responds to both external and internal stimuli and transmits the impulses to other parts of the body, such as muscles and glands, so that the body responds to the environment. It should be emphasized that the body need not necessarily produce a response to a stimulus (inhibition).

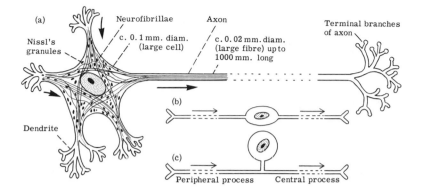

Fig.17. (a) Multipolar neurone, (b) Bipolar neurone, (c) Pseudo-unipolar neurone.

THE NEURONE. The cell which especially possesses this property of irritability and of being able to transmit an impulse is called a *neurone*. It consists typically of a *cell body* and *processes* (Fig.17a). In one type of neurone, there are many, relatively short, processes which are thick and branched (*dendrites*) and one long process (the *axon*) which has a few branches at right angles to it at its beginning but may be regarded as unbranched until it divides at its termination. One of the important features of the axon is that it can be very long (over 1 metre) although in some parts of the nervous system the axon may be only a few millimetres long. Some measurements may indicate the peculiar structure of the neurone. The cell body, if very large, may be up to 100 μm in diameter. The axon, if very large, may be about 20 μm in diameter, and that includes one of its coverings. One therefore has to think of a process which may be 50,000 times as long as it is wide. One can see how a cell with this structure is eminently suitable for linking together different parts of the body.

TYPES OF NEURONE. The type of neurone just described is called *multipolar*. There is however another type in which there are only two processes which are very much alike (*bipolar neurone*) (Fig.17b). Since conduction in a multipolar neurone is usually from the dendrites to the cell body to the axon, the process of a bipolar cell which conducts an impulse to the cell may be called the dendrite and the other the axon. However, this type of cell is associated with the conduction of impulses from the periphery of the body.

The dendrite is therefore called the *peripheral process*. The axon is called the *central process* because it conducts the impulse from the cell towards the aggregation of nervous tissue called the spinal cord and brain (the *central nervous system*). Usually a bipolar neurone is modified in the course of development and forms a *pseudo-unipolar neurone* in which one process leaves the cell body and divides into two, one peripheral and one central. Bipolar neurones are found in the retina of the eye and elsewhere.

THE STRUCTURE OF A NEURONE. The nucleus of the cell body of a neurone is usually large with comparatively little chromatin and a conspicuous nucleolus. The cytoplasm of the cell contains granules which stain with basic dyes (for example, methylene blue). These are called *Nissl's granules* and are found only in neurones. There are also fine threads which stain with silver nitrate and are called *neurofibrillae*. These however in spite of their name are not found only in nerve cells. There are also mitochondria, a Golgi apparatus and frequently pigment in the cell body.

The Nissl's granules extend into the dendrites for a short distance but there are no granules in the axon. Both types of processes contain neurofibrillae. The typical multipolar cell has a large number of dendrites, each with a large number of branches. The axon on the other hand is unbranched and finally divides and subdivides a large number of times. The terminations of the fine fibrils thus formed are often slightly enlarged. They are called *boutons terminaux* or *end feet* (Fig.18). Neurones are so arranged that they can transfer an impulse to one another. This is done by means of the close relationship between these terminal boutons of an axon, and the dendrites and the cell body of another neurone. This close relationship is called a *synapse* (a term which is the Greek form of the Latin word *contact*). There is no continuity between the processes of one neurone and those of another. At the synapse an impulse can travel in only one direction, from the terminal boutons of one neurone to the dendrites and/or cell body of another. It is important to realize that through its terminal divisions one axon makes contact with a large number of neurones, and that the dendrites and cell body of one neurone receive the end feet of the axons of a large number of neurones.

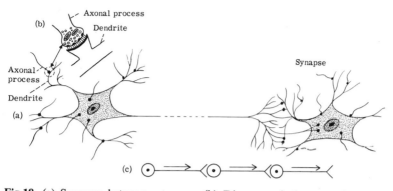

Fig.18. (a) Synapses between neurones, (b) Diagram of electron microscope appearance of one type of synapse, (c) Conventional diagrammatic representation of a chain of neurones.

FUNCTIONAL TYPES OF NEURONE. Neurones are also classified in terms of their function. If they conduct an impulse from the periphery of the body towards the central nervous system they are called *sensory* (*afferent* or *receptor*) neurones. Within the body there may be one or more neurones conducting the impulse to the final neurone whose axon ends in relation to the muscle or gland which contracts or secretes as a result of the original stimulus. The former are called *connector* (*intercalated*) neurones and the latter *motor* (*efferent* or *effector*) neurones.

THE STRUCTURE OF NERVE FIBRES. Axons consist of fluid (*axoplasm*) surrounded by a membrane, the *axolemma*, and are surrounded by coverings or sheaths except at their beginnings and terminations. (Axons are usually referred to as nerve fibres.) The smallest axons have very much thinner sheaths than the larger axons although the relationship between the two is not constant. Both the peripheral and central processes of a pseudo-unipolar or bipolar cell have sheaths. The dendrites of multipolar cells have no sheaths.

The sheath immediately surrounding the axon is called the *myelin sheath* (Fig.19). It consists of a variable number of lamellae of alternate fat and protein. Large axons usually have a thick myelin sheath and very small axons have a myelin sheath consisting of only one or two lamellae. It is the myelin sheaths of nerve fibres which make them look white. Hence the term *white matter* to describe bundles of nerve fibres in the central nervous system. Peripheral

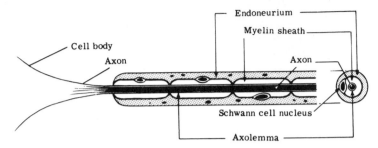

Fig.19. A longitudinal section, and a transverse section of a peripheral nerve fibre and its sheaths.

nerve trunks, or simply nerves, similarly look white. The standard staining techniques do not show very thin myelin. Because of this they are called *non-myelinated nerve fibres*. On the other hand those fibres with easily stained myelin sheaths are called *myelinated*. Myelinated fibres can be large or small. There is broadly speaking a relationship between the speed of conduction of the impulse and the size of the fibre (and also the degree of myelination). The larger the fibre, the faster is the speed of conduction (the fastest speed is about 100 metres per second and the slowest about 1 metre per second).

The myelin sheath is interrupted at intervals. These interruptions are called the *nodes of Ranvier*. Round the myelin sheath is a covering called the *Schwann cell*. There is one Schwann cell round each internodal segment (the part between two nodes). The myelin sheath is formed as a result of an interaction between the axon and the Schwann cell in which the Schwann cell spirals round the axon. The myelin consists of Schwann cell membrane. Outside the Schwann cell is a connective tissue sheath called the *endoneurium*. The endoneurium dips down and meets the axon at the nodes of Ranvier. Within the brain and spinal cord there is no connective tissue endoneurium. Instead, the myelinated nerve fibres are related to cells called *oligodendrocytes* (similar to Schwann cells) and a network of fibres which are the processes of what are known as *neuroglial cells*, which include the oligodendrocytes.

THE SPINAL NERVES. Reference has been made to the central nervous system consisting of the brain and spinal cord. The nerve trunks (or simply nerves) of the body contain fibres which convey impulses from the different parts of the body to the central nervous system (that is, sensory fibres) and impulses from the central nervous

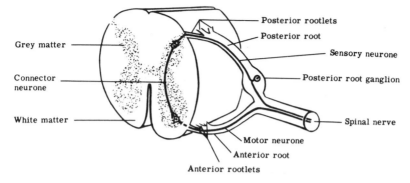

Fig.20. The formation of a spinal nerve.

system to the muscles of the body (that is, motor fibres). These nerves form the *peripheral nervous system*. They enter or emerge from the brain inside the cranium, or enter or emerge from the spinal cord in the vertebral canal and constitute the twelve pairs of *cranial nerves* and thirty-one pairs of *spinal nerves* respectively. It can be seen that a spinal nerve is a *mixed nerve*, that is, it contains sensory and motor fibres. As one approaches the spinal cord the spinal nerve separates into a *sensory root* and a *motor root* each of which subdivides into *rootlets* (Fig.20). The cell bodies of the sensory fibres are in an enlargement of the posterior root (*posterior root ganglion*) and the cell bodies of the motor fibres are in the spinal cord itself. The spinal cord if examined in transverse section is seen to be oval and consist of a central fluted column of *grey matter* like the letter H, surrounded by columns of *white matter*. Aggregations of cell bodies of neurons are called grey matter and aggregations of nerve fibres are called white matter because this difference in colour can actually be seen. The whiteness of white matter is due to the myelin sheaths of the nerve fibres.

There are eight pairs of cervical spinal nerves (the first pair emerge between the occipital bone and the atlas and the eighth between the seventh cervical and the first thoracic vertebra), twelve thoracic, five lumbar, five sacral and one coccygeal. A spinal nerve emerges from an intervertebral foramen and immediately divides into an *anterior primary ramus* (ramus = branch) and *posterior primary ramus*. All the posterior primary rami are distributed to the muscles on either side of the vertebral column and to about 5 cm of the skin on either side of the midline of the back. The much larger anterior primary rami usually form *plexuses, cervical* in the upper part of the neck, *brachial* in the lower part of the neck, *lumbar* on the posterior abdominal wall and *sacral* in the pelvis. These plexuses or networks finally form branches which are mixed nerves and are distributed to the muscles and the skin of the body.

3

The locomotor system. I

This system consists of the skeleton, joints and muscles. The skeleton and muscles more or less give the body its shape, although fat between the skin and the underlying muscle alter the contours of the body, and the contents of a cavity, such as the thoracic and abdominal cavities, influence the shape of the thorax and the abdomen. The presence of joints, defined simply as the various unions between bones, allows one bone to move on another and these movements are brought about by muscles. The contraction of muscles and the co-ordination of movements are examples of how impossible it is to separate one system from another. Properly functioning vascular and nervous systems are essential for correct muscle function.

The skeleton (Fig.21)

A knowledge of the human skeleton makes it possible to understand more easily the arrangement of the rest of the tissues and organs of the body since the bones or parts of the bones are frequently used as points or areas of reference. The skeleton provides the body with rigidity, allows movement because of its many parts, affords protection in many sites, is a storehouse of calcium and has blood-forming functions. The skeleton is usually divided into the *axial* (vertebrae, skull, ribs and sternum) and the *appendicular* (pectoral and pelvic girdles and limb bones).

THE VERTEBRAL COLUMN. The *vertebrae* in man usually number 33 which are divided into 7 *cervical*, 12 *thoracic* (*dorsal*), 5 *lumbar*, 5 *sacral* (forming one bone, the *sacrum*), and 4 *coccygeal* (forming the *coccyx*) (Fig.22). A typical vertebra has an anterior cylindrical

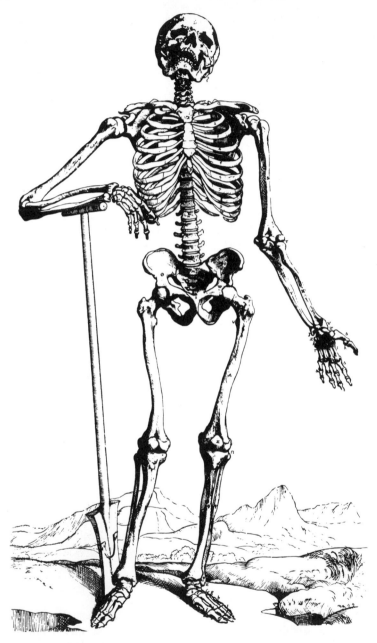

Fig.21. The human skeleton, from Vesalius' *De Humani Corporis Fabrica* (Concerning the Structure of the Human Body).

part, the *body*, which becomes larger from above downwards. This is associated with the transmission of the weight of the body to the pelvis and lower limbs. Between the bodies are *intervertebral discs* of fibro-cartilage which are firmly attached to the adjacent bodies' surfaces. Behind the body is the *vertebral arch* which together with the back of the body encloses the *vertebral canal* (Fig.23a).

The part of the vertebral arch passing backwards and laterally from the upper part of each side of the back of the body is called the *pedicle*. It is usually rounded. The posterior part of the arch which passes backwards and medially on each side is called the *lamina*. It is usually flattened. Projecting backwards from the back of the vertebral arch, in the midline, is the *spinous process* (*spine*), projecting laterally, one on each side are the *transverse processes* and projecting upwards and downwards from the sides of the arch at the junction of the pedicles and laminae are the two *superior articular processes* and the two *inferior articular processes* (Fig. 23c). Usually the front of the inferior articular process of the vertebra above articulates with the back of the superior process of the vertebra below. A *costal process* also develops laterally in relation to the side of the body and the front of the transverse process but only in the thoracic region does it become a *rib*. Normally there are no ribs in the other regions of the axial skeleton.

When looked at from the side, an opening between two adjacent vertebrae can be seen. It is bounded by the bodies in front, by the pedicles above and below and the articulation between the articular processes

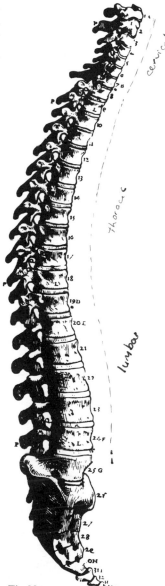

Fig.22.
The vertebral column, from Vesalius' *De Humani Corporis Fabrica* (Concerning the Structure of the Human Body).

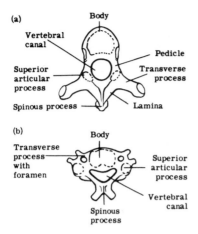

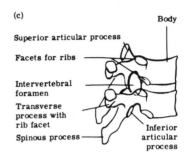

Fig.23. (a) Thoracic vertebra from above, (b) Cervical vertebra from above, (c) Two articulating thoracic vertebrae from the side to show the formation of an intervertebral foramen.

behind. This is called the *intervertebral foramen*. Through it runs a spinal nerve (Fig.23c).

An articulated vertebral column if looked at from the side has a forward curve in the cervical region, a backward curve in the thoracic, and a forward curve in the lumbar. In a fetus *in utero* there is only a backward curve for the whole of the vertebral column. The forward curve of the neck region is associated with the child's holding its head up at about six weeks, and the forward curve of the lumbar region with the child's sitting up and standing at about six months to a year.

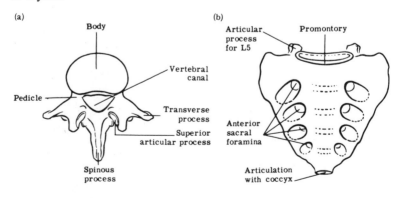

Fig.24. (a) Lumbar vertebra from above, (b) Sacrum from in front.

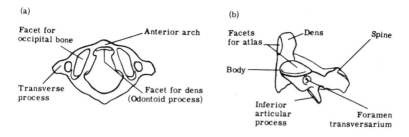

Fig.25. (a) Atlas from above, (b) Axis from the side.

A cervical vertebra is distinguishable because its transverse process has a hole (foramen) in it (*foramen transversarium*). Associated with the articulations of the ribs with the bodies of the thoracic vertebrae, these are facets on the bodies of the thoracic vertebrae and thus thoracic vertebrae can be identified. Lumbar vertebrae have neither foramina transversaria nor facets for the ribs (Figs.23b, 24a).

There are other differences between the vertebrae such as the increasing size of the body from above downwards already referred to. The spine varies. It is bifid in most of the cervical vertebrae, long and downward projecting in the thoracic and square and horizontal in the lumbar.

The first cervical vertebra is known as the *atlas* and consists of a ring of bone without a body (Fig.25a). It articulates above with the base of the skull and below with the second cervical vertebra, called the *axis* (Fig.25b). The body of the axis projects upwards to form the *dens* (*odontoid process*) which represents the body of the atlas.

The sacrum is triangular in shape with the apex below (Fig.24b). On either side it has a wide area for articulation with the hip bone. Below the articulation it tapers fairly quickly. It is easy to identify various parts of individual vertebrae in the sacrum but the coccyx shows few of these.

The vertebral canal contains the spinal cord and its coverings (*meninges*), the origins of the spinal nerves, blood vessels and some fat. The spinal cord extends downwards as far as the lower border of the first lumbar vertebra but the roots of the spinal nerves continue downwards to the lowest part of the vertebral canal.

The bodies of the vertebrae consist of an outer covering of compact bone and inner cancellous bone in the meshes of which there is red bone marrow. The bodies of the vertebrae form a very important site for the formation of blood cells. Associated with this is a large opening on the back of the body of a vertebra, especially the lumbar vertebrae, out of which emerges a large vein which joins a venous plexus in the vertebral canal.

THE SKULL. The *skull* consists of the *cranium* (the head and face) and the *mandible* (lower jaw). The cranium has an upper box-like part containing the brain and a front lower part consisting of the facial skeleton. All the bones of the skull are firmly united to each other in the adult except for the mandible, which is fairly freely movable. The skull may be regarded as consisting of a number of cavities. The brain lies in the largest of these. Two cavities, one on each side of the skull above and in front, are called the *orbits* and contain the eyeball and its muscles, vessels and nerves. Between and below the orbits is the *nasal cavity*, divided into two by a vertical septum. This cavity constitutes the beginning of the *respiratory system* but also contains the nerves associated with the sense of smell (*olfaction*). Below the nasal cavity is the *mouth* although this is not seen as a cavity of the skull itself. The mouth is the beginning of the *alimentary system*, communicates with the respiratory system and contains the nerves associated with the sense of *taste*. Many of the bones of the skull have air-containing cavities communicating with the nasal cavities and the *ear* lies in a cavity in a bone (*temporal*) in the base of the skull.

The skull from in front. The main bones which can be seen from the front of the skull (Fig.26a) are the *frontal bone* above, with a vertical part forming the forehead and a horizontal part dividing into two and forming the roof of each orbit, the *zygomatic* (cheek) *bones* forming the lateral wall of the orbit and articulating above with the frontal bone of each side, and the *maxillae*, one on each side of the midline below the zygomatic bone. They form the floor of the orbit and the lateral boundary of the nasal opening, and articulate with the frontal bone above to complete the margin of the orbit. Medial to the upper part of the maxillae are the *nasal bones* which meet in the midline and complete with the maxillae the margin of the nasal opening. Below the nasal opening the two maxillae meet and complete the upper alveolar arch containing the teeth.

The skull from the side. From the side, the upper part of the cranium (the *vault* of the skull) comprises from in front backwards the *frontal, parietal* and *occipital bones* (Fig.26b). There are two parietal bones meeting in the midline above in the *sagittal suture*. The transverse articulation between the frontal bone and parietal

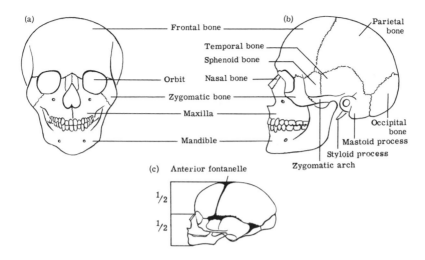

Fig. 26. (a) Skull from in front, (b) Skull from the side, (c) Infant skull from the side.

bones is called the *coronal suture* and that between the parietal and occipital bones the *lambdoid suture*. The occipital bone extends into the base of the skull behind. In front of the occipital bone and below the parietal bone is part of the *temporal bone* which also extends into the base of the skull in front of the occipital bone. The *sphenoid bone* lies in front of the temporal bone and extends into the base of the skull in front of the temporal bone. The temporal bone articulates above with the parietal bone and behind with the occipital. The sphenoid bone on the side of the skull articulates with the parietal bone above and the zygomatic bone in front. The latter bone and the maxilla can be seen on the side of the skull in front. There is a forward projection from the temporal bone just in front of the opening for the ear (*external auditory meatus*). This meets the zygomatic bone and forms the *zygomatic arch*. This arch is easily felt in the living subject and if followed forward leads to the zygomatic bone forming the projection of the cheek. Between the arch and the surface of the skull is the *temporal fossa*. Behind and below the opening for the ear is a downwardly projecting piece of bone (*mastoid process*) which is part of the temporal bone.

The skull from below. On the inferior aspect of the skull (Fig. 27a) the occipital bone is seen at the back. In it, in the midline, is the large *foramen magnum*, a communication between the inside of the skull

(a)   (b)

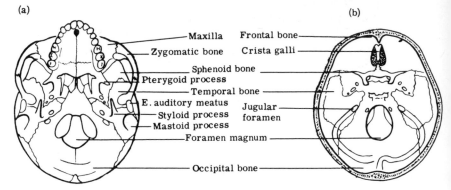

Fig.27. (a) Base of skull from below, (b) Base of skull from above.

and the vertebral canal. On either side of the foramen magnum
nearer the front are the *occipital condyles* which articulate with the
upper surfaces of the atlas. In front of the foramen magnum a
narrow part of the occipital bone fuses with the *body* of the sphenoid
bone which extends laterally and forwards on both sides to where it
bends upwards at a right angle to form part of the temporal fossa
seen deep to the zygomatic arch. Laterally between the occipital bone
behind and the sphenoid in front and articulating with both is the
petrous part of the temporal bone passing medially and forwards.
It is continuous with the mastoid process. In front of the external
auditory meatus is a hollowed-out area called the *mandibular fossa*.
The *jugular foramen* is an opening between the temporal and occipital
bones. Projecting downwards and forwards from the temporal bone
medial to the mastoid process is the *styloid process* which is about
2.5 cm long and 3 mm wide. In the midline in front of the body of
the sphenoid is the vertically placed *vomer* forming part of the septum
of the nasal cavity. The lateral boundary of the nasal cavity is a
downwardly projecting part of the sphenoid bone called the *pterygoid process* which has two backwardly projecting plates of bone,
the *lateral* and *medial pterygoid plates*. The floor of the nasal cavities
(*palate*) is formed mainly by the maxillae and to some extent behind
by the *palatine bones*. The hard palate is bounded by the alveolar
arch formed by the maxillae. Foramina can be seen in the midline
of the palate near the front and at each posterolateral angle.

The inside of the skull. The interior of the base of the skull consists
of the frontal bone in front, the sphenoid bone in the middle and the
occipital bone behind (Fig.27b). The temporal bone is wedged
laterally between the sphenoid and occipital bones. There is a gap

in the frontal bone in the midline and in the gap are the *cribriform plates* and *crista galli* of the *ethmoid bone*. There are many small holes in the horizontal cribriform plates. The crista galli is in the midline and consists of a projecting piece of bone flattened from side to side. The upper surface of the body of the sphenoid is hollowed out and forms the *sella turcica*. A sagittal groove can be seen in or near the midline of the vault of the skull passing backwards from the crista galli and lying in the frontal, parietal and occipital bones. At the posterior end of the vault about 5 cm above the foramen magnum this *sagittal sulcus* turns to the right as a rule and becomes the *transverse sulcus* on the occipital bone. There is a transverse sulcus on the left which is continuous with a sulcus passing upwards in the midline from the foramen magnum. The transverse sulcus turns downwards and medially, and grooves the temporal bone. It ends at the jugular foramen.

The mandible. The mandible forms the skeleton of the lower jaw and in it are the lower teeth (Fig.28a,b). It consists of a horizontal, anterior horseshoe-shaped part, the *body*, and on each side a posterior vertical part, the *ramus*. The body has internal and external surfaces and upper and lower borders. In the midline in front is the *symphysis menti* where the two halves, from which the mandible developed, are united. There is a protuberance on the lower half of the external surface of the symphysis. About 2.5 cm from the symphysis on either

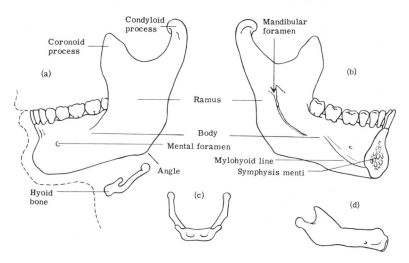

Fig.28. (a) Outside of mandible, (b) Inside of mandible, (c) Hyoid bone, (d) Infant's mandible.

side is an opening, the *mental foramen*, midway between the upper
and lower borders. On the internal surface running somewhat
downwards and forwards is the *mylohyoid line*. Above this line the
body of the mandible is in the floor of the mouth; below it, the body
lies in the neck. The ramus is quadrangular in shape. The *angle* of
the mandible is where the posterior border meets the inferior border.
There are two upward projections on the superior border. The
anterior is called the *coronoid process* which is triangular with the
apex pointing upwards. It lies deep to the zygomatic arch when the
mouth is closed. The posterior is called the *condyloid process* and is
enlarged transversely above to form the *head* which articulates with
the mandibular fossa in the base of the skull. If a finger is placed
in the external auditory meatus the head can be felt moving forwards
and backwards when the mouth is opened and closed. Near the
middle of the internal surface of the ramus there is an opening called
the *mandibular foramen* leading to the *mandibular canal* in which
lies the nerve supplying the lower teeth.

The hyoid bone. The hyoid bone is often described with the skull.
It can be felt in the uppermost part of the neck just below the level
of the lower border of the mandible. It consists of a *body* in the
middle (about 4–5 cm wide and 1 cm high), two *greater horns* one
on each side projecting backwards from the body and two *lesser
horns* projecting upwards and backwards from the junction of the
body with the greater horns (Fig.28c).

The growth of the skull. Broadly speaking, the vault of the skull
ossifies in membrane and the base in cartilage. The main face bones
ossify in membrane. At birth the head forms a much greater proportion of the total body length than in the adult (about one quarter as
compared with an eighth). Within the skull itself there are also
marked differences. In an infant the facial skeleton is a much smaller
proportion of the front of the head than in the adult. From the upper
border of the orbit to the lower edge of the jaw is about one-half of
the height of the skull in the newborn (Fig.26c). In an adult it is
about two-thirds. This change is due to the growth of the maxilla
and mandible associated with the eruption of the teeth, and the
growth of the space in the maxilla and enlargement of the nasal
cavity. The orbits are much larger proportionately at birth than in the
adult.

At birth the bones of the vault of the skull are separated by fibrous
tissue and their edges are not serrated. Growth of the skull takes
place at the edges. There is also growth by the addition of bone to

the outside of the skull and removal from the inside. There are large areas of fibrous tissue in certain places especially in the sagittal suture where the frontal bone meets the parietal bones. This is diamond-shaped and is called the *anterior fontanelle* (Figs.26c, 33c). It closes between eighteen months and two years. The growth of the skull is most rapid in the first year of life and stops at about sixteen years.

The temporal bone shows marked changes in a child's early years. At birth the external auditory meatus is shallow, so that the tympanic membrane (ear-drum) is near the surface, and the mastoid process has not yet developed. These bones grow in the first five years of life.

There is some growth of the maxilla and mandible during the eruption of the milk teeth but growth is much more marked while the permanent teeth are appearing between six and fourteen years of age. At birth the angle between the ramus and body of the mandible is obtuse and in the adult it is nearly a right angle (Fig.28a,d).

THE THORACIC CAGE. The skeleton of the *thorax* consists of the thoracic vertebrae and intervertebral discs behind, the ribs posteriorly, laterally and anteriorly and the *costal cartilages* and *sternum* in front. (Fig.29). The sternum is in the midline in front and the costal cartilages extend from the anterior ends of the ribs to the sternum, as a rule. The sternum is a flattened bone about 12.5 cm long and 4 cm wide. It consists of an upper part, the *manubrium* (about one-third of its length), a middle part, the *body*, and a small lower part, the *xiphoid process* (Fig.29). The manubrium is roughly quadrilateral in shape and the medial ends of the clavicle (collar bone) and first costal cartilage articulate with its upper lateral angle. The manubrium articulates below with the body at the *manubriosternal joint* and the transverse line of articulation can be

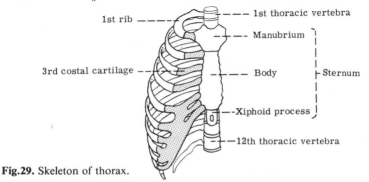

Fig.29. Skeleton of thorax.

felt as a projecting ridge. This is called the *sternal angle*. The second costal cartilage articulates with the sternum opposite the manubriosternal joint. The lowest costal cartilage to articulate with the sternum is the seventh. The sternum is a good example of a flat bone and consists of compact bone outside and cancellous bone on the inside. Since it is just under the skin it is used for obtaining samples of red bone marrow (*sternal puncture*).

There are twelve ribs on each side. They form a series of obliquely placed bony arches. The first rib is less oblique than the others and the eighth is the most oblique. Each rib articulates posteriorly by means of its *head* with the bodies of two neighbouring vertebrae (Fig.43a) and anteriorly with its costal cartilage through which it articulates with the sternum (first to seventh costal cartilage) or with the costal cartilage above (eighth with seventh, ninth with eighth, tenth with ninth). The exceptions are the costal cartilages of the eleventh and twelfth ribs which end freely in the abdominal wall (*floating ribs*). The first to the seventh ribs are called *true* and the eighth to the tenth *false*. Next to the head of a rib is a narrow part called the *neck* and then there is a projection posteriorly called the *tubercle*. About 5–6 cm lateral to the tubercle a typical rib suddenly bends forwards at the *angle* and forms the *shaft* which is flattened from side to side so that the rib has external and internal surfaces and upper and lower borders. The *costal groove* is seen on the lower half of the internal surface of the shaft. The anterior end of a rib articulates with a costal cartilage. The medial articulations of the costal cartilages with the sternum or with each other have already been considered.

The thorax as a whole is roughly conical in shape with a relatively small *inlet* above and a much larger *outlet* below (Fig.29). The inlet is kidney-shaped and is bounded by the first thoracic vertebra behind, the first rib laterally and the manubrium in front. It is oblique so that anteriorly it is at the level of the second thoracic vertebra. The inferior opening (outlet) of the thorax is bounded by the twelfth thoracic vertebra, the twelfth and eleventh ribs and the costal cartilages of the tenth to the seventh ribs. The plane of the outlet slopes upwards from behind and its edge is called the *costal margin*. The vertebral bodies project forwards in the midline at the back of the thorax and thus narrow the anteroposterior diameter in this part. There is, however, a marked recess on either side of the vertebrae in front of the posterior ends of the ribs where they pass backwards in front of the transverse processes (Fig.45).

THE UPPER LIMB BONES. The upper appendicular skeleton consists of the *pectoral girdle* and the upper limb bones. The pectoral girdle consists in front of the tubular *clavicle* (collar bone) which lies transversely and the triangular *scapula* which lies on the back part of the upper seven ribs (Fig.30a,b). Projecting horizontally backwards is the *spine* of the scapula which if followed laterally is seen to turn forwards to form a shelf known as the *acromion*. The clavicle articulates laterally with the medial border of the acromion and medially with the manubrium of the sternum and the first costal cartilage. The upper lateral angle of the scapula has a shallow saucer-like area called the *glenoid cavity*. Medial to its upper part and projecting forwards below the acromion is the *coracoid process*.

The *humerus* is the bone of the upper arm (Fig.30c). It has a more or less rounded shaft and an upper end which has a hemispherical *head* facing medially for articulation with the glenoid cavity at the shoulder joint. The distal end of the humerus is flattened and has a lateral articular area (*capitulum*) which is rounded, and a medial articular area (*trochlea*) which is pulley-shaped. The lateral bone of the forearm is called the *radius* (Fig.30d). Its upper end has a rounded *head*, which proximally articulates with the capitulum, and a distal end which has an articular area on it. The radius is more or less tubular as is the *ulna*, the medial bone of the forearm. The upper end of the ulna has an anterior notch into which fits the trochlea of the humerus. This articulation together with the radiohumeral articulation forms the *elbow joint*. The radial head medially articulates with a notch on the lateral side of the upper end of the ulna. The lower end of the ulna does not have an articular area on its distal surface but its lateral side articulates with a notch on the medial side of the lower end of the radius (Fig.30d).

The *carpus* is the name given to the bones of the wrist. There are eight bones in the carpus, in two rows of four. The proximal row articulates superiorly with the radius laterally at the wrist joint and a disc of fibrocartilage medially. This disc separates the ulna from the wrist joint. The distal row of bones of the carpus articulates with the five *metacarpal bones*. (The medial bone of the distal row articulates with two metacarpal bones.) The distal end of the metacarpal bone forms a knuckle and a digit consists of three *phalanges* (*proximal*, *middle* and *distal*) except for the thumb which has only two phalanges. There is considerable movement between the first (thumb) metacarpal and the carpus but little movement at the articulations between the carpus and the other metacarpals (Fig.30e).

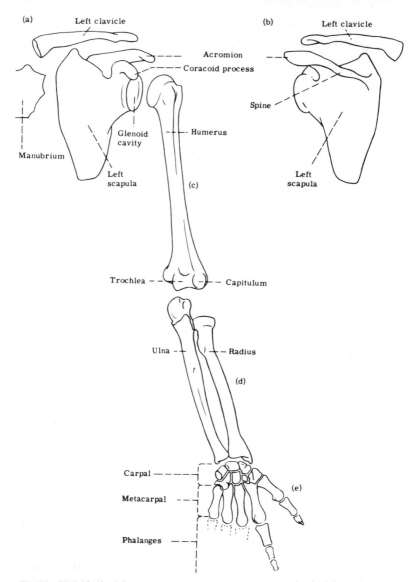

Fig.30. (a) Left clavicle and scapula from in front, (b) Left clavicle and scapula from behind, (c) Left humerus, (d) Left radius and ulna, (e) Left carpal, metacarpal bones and phalanges.

THE LOWER LIMB BONES. The lower appendicular skeleton consists of the *pelvic girdle* and the lower limb bones. The pelvic girdle (*pelvis*) consists of the two *hip bones* and the sacrum (Fig.31). The sacrum lies posteriorly between the two hip bones and articulates with the hip bones at the *sacro-iliac joints*. The hip bone consists of three parts, the *ilium* above, the *ischium* below and behind and the *pubis* below and in front. These bones meet in the region of the *acetabulum*, a large cup-shaped depression on the lateral side of the hip bone. The ilium is a large flat bone with an upper border extending anteroposteriorly, the *iliac crest*. The anterior part of the two pubic bones meet in the midline at the *symphysis pubis*. Below the acetabulum is the large *obturator foramen*, and behind this is the *ischial tuberosity*, a rather massive piece of bone on which we sit (Fig.32a,b,c).

The inside of the pelvis is divided into the *false* (above the level of an oblique plane through the upper edge of the sacrum and upper edge of the symphysis) and the *true* (below this plane) (Fig.31). The false pelvis forms part of the abdominal cavity. The importance of the true pelvis in the female is that it forms a bony canal through which the head of the baby has to go during childbirth. Associated with this is the fact that the true pelvis in the female is generally wider and shallower than in the male (Fig.31a,b). This results in important differences between the male and female pelvis and makes it possible to establish the sex of a skeleton. The small diagrams indicate the main differences in the shape of the inlet, cavity and outlet of the pelvis in the male and female. At the inlet the main

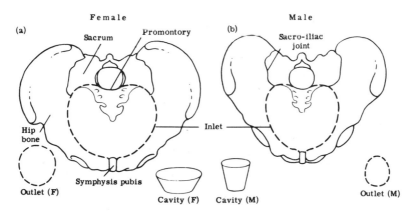

Fig.31. (a) Female pelvis, (b) Male pelvis. (F=female; M=male)

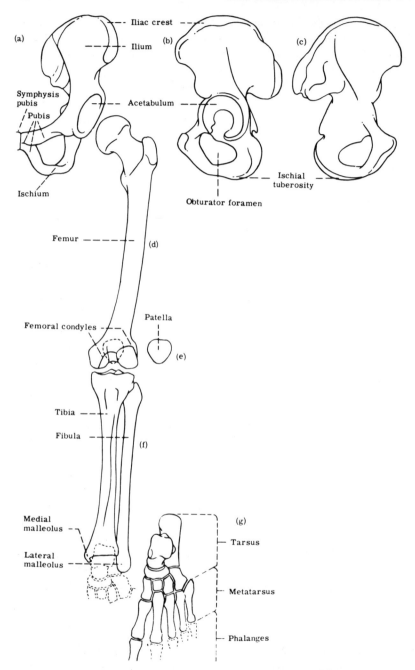

Fig.32. (a) Left hip bone from in front, (b) Left hip bone from outside, (c) Left hip bone from inside, (d) Left femur, (e) Patella, (f) Left tibia and fibula, (g) Bones of left foot.

difference is in the transverse diameter which is smaller in the male than in the female. The cavity in the female pelvis is much roomier and at the outlet all the diameters of the male pelvis are smaller than those in the female.

The *femur* is the bone of the thigh (Fig.32d). It is a long tubular bone in which the upper end turns medially and articulates, by means of a spherical *head*, with the acetabulum at the hip joint. The lower end is enlarged to form two prominences called the *condyles*. These structures are articular both distally and anteriorly. The central anterior articular area is for the *patella* (knee-cap). The bones of the leg are called the *tibia*, the medial and more massive, and *fibula*, the lateral and more slender (Fig.32f). The upper end of the tibia is enlarged and forms two *condyles*, and on its upper surface are two articular areas which together with the articular areas on the distal surface of the condyles of the femur form the knee joint.

The distal ends of the tibia and fibula project medially and laterally and can be felt in the region of the ankle. The tibial projection is called the *medial malleolus* and the fibular, the *lateral malleolus*. The lower ends of the tibia and fibula are held together by fibrous tissue above the ankle. The bones of the foot are called the *tarsus*, the *metatarsus*, and the *phalanges* from proximal to distal (Fig.32g). The uppermost bone, the *talus*, lies below and between the tibia and fibula and the articulation between these three bones is the ankle joint. The malleoli grip the talus between them. Below the talus is the *calcaneus* which projects backwards to form the heel. Between the talus and calcaneus, and the five metatarsal bones there are five more tarsal bones and four of them form a row with which the metatarsal bones articulate. The fourth and fifth metatarsals articulate with one tarsal bone. Each metatarsal bone articulates with a proximal phalanx. There are three *phalanges* in each toe except for the first (big) toe which has only two. There is very little movement between the tarsal bones except for the quite extensive movement between the talus and the rest of the foot. There is little or no movement between the tarsus and metatarsus but the toes can be turned upwards and downwards between the metatarsus and phalanges (particularly upwards) and between the phalanges (particularly downwards).

Joints

A joint may be defined as the union between two bones and the way in which the bones are held together determines the type of joint.

Bones may be held together by fibrous tissue between their ends (*fibrous joints*), or by cartilage (*cartilaginous joints*), or they may remain separate and be held together by a ligamentous capsule lined by synovial membrane (*synovial* or *diarthrodial joints*).

FIBROUS JOINTS. Fibrous joints are found in the skull between the bones in the vault and between the bones of the face. They are called *sutures*. In the vault the edges are frequently serrated and fit into one another (for example, *sagittal* and *coronal sutures*) (Fig.33a,b,c). These joints disappear in time and are replaced by bone (*synostosis*) and this is one way of estimating the age of a skull because they disappear from the inside of the skull first and usually in a certain order. It should be noted that while the joint is present the fibrous tissue between the bone is continuous with the periosteum of the bones forming the joint. Another example of a fibrous joint is the union between the adjacent lower ends of the tibia and fibula above the ankle joint (Fig.33d). A tooth is held in its socket by fibrous tissue and this is also referred to as a fibrous joint (Fig.33e).

It is obvious that there is usually no movement at this type of joint.

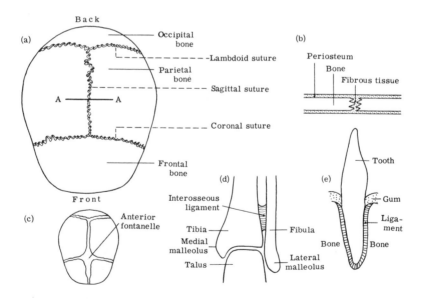

Fig.33. Fibrous joints: (a) Sutures of skull, (b) Section through suture at A-A, (c) Sutures in infant's skull, (d) Joint between lower ends of tibia and fibula, (e) Articulation of tooth with jaw.

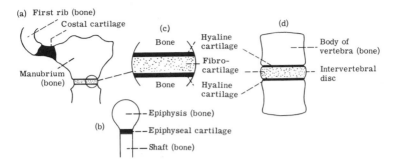

Fig.34. Cartilaginous joints: (a),(b) Primary, (c),(d) Secondary.

If, however, there is a great deal of fibrous tissue between the bones, movement can take place. This is seen in the sutures of the skull of a full-time fetus and during childbirth the bones are capable of moving so that one lies underneath another. This is called *moulding* and reduces the size of the baby's head.

CARTILAGINOUS JOINTS. Cartilaginous joints are of two types, *primary* and *secondary*. In a primary cartilaginous joint the two bones forming the joint are united by hyaline cartilage (Fig.34a,b). In the adult this is seen only at the junction of the first rib with the manubrium of the sternum. In the growing child there are many places where two pieces of bone are joined by hyaline cartilage, for example, the epiphysis with the diaphysis in a long bone. All these disappear when growth is complete.

In secondary cartilaginous joints the bone ends are covered by hyaline cartilage and between them is a disc of fibrocartilage (Fig.34c,d). It is usually said that all these joints are found in the midline of the body. They are

    a. the joints between the bodies of the vertebrae in which the adjacent surfaces of the bodies are covered by hyaline cartilage and between them is the intervertebral disc,

    b. the manubriosternal joint,

    c. the symphysis pubis between the anterior ends of the pubic bones.

All these joints have considerable functional significance. The joints between the bodies of the vertebrae permit movement by compressing one side of the disc and pulling apart the opposite side. The movement is obviously limited in extent but the amount of movement between each pair of vertebrae is summated so that there is considerable forward bending (*flexion*) of the whole vertebral column. The opposite movement (*extension*) is more limited due to the spinous processes of the thoracic vertebrae coming together but very extensive movement in the lumbar region can be seen in some acrobats. Excessive bending forwards can tear the back of an intervertebral disc so that its somewhat softer centre oozes backwards and may press on the roots of the spinal nerves in the lower part of the vertebral canal.

There is also some movement at the manubriosternal joint during respiration. On respiration the body of the sternum moves up and increases the forward angulation at this joint. After the age of 40 years the joint frequently becomes synostosed and movement at this joint disappears.

Normally there is no movement at the symphysis pubis but in pregnant women the ligaments around the joint become softened and the bone ends may even become absorbed. If this proceeds too far the woman may complain of a feeling of instability while walking. The normal state of affairs returns after childbirth.

SYNOVIAL JOINTS. In a synovial joint the bones are separate and their ends are covered by hyaline (*articular*) cartilage. The bones are held together by a *capsule* (*capsular ligament*). This is usually attached to the ends of the bone just beyond the articular cartilage. The capsule is lined by *synovial membrane* which always ends at the articular cartilage (Fig.35a). The potential cavity of the joint is therefore surrounded by synovial membrane and the articular cartilage.

If the capsule is attached beyond the edge of this cartilage the synovial membrane leaves the capsule and covers the bone as far as the cartilage (Fig.35b). The capsule may be thickened in places. These thickenings form *intrinsic ligaments*. If however, there are bands or sheets of fibrous (collagenous, ligamentous) tissue between the bones forming the joint and separate from the capsule, they are called *extrinsic ligaments*. Sometimes the synovial membrane protrudes beyond the capsule and sometimes there are bursae near the joint.

In some joints there are structures inside the capsule but outside the synovial membrane. They are referred to as *intra-articular structures*. They may be cartilaginous discs or ligaments or tendons but a much better term would be *intracapsular* (Fig.35c,d,e). Cartila-

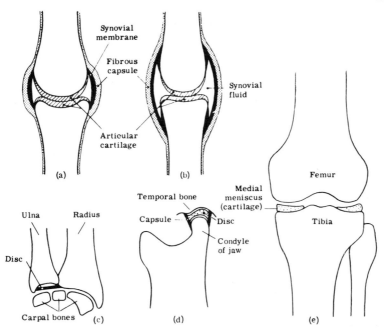

Fig.35. (a) Section through typical synovial joint, (b) Section through synovial joint in which capsule is attached beyond articular cartilage, (c, d, e) Wrist, mandibular and knee joints showing intra-articular fibrocartilaginous discs (menisci in knee joint).

ginous discs are found, for example, in the knee, mandibular, sternoclavicular and wrist joints.

The ligaments, capsular and others, are very strong and hold the bones together. The direction of their fibres also influences the way in which one bone moves on the other and the looseness of the capsule is also related to the movement which takes place at the joint. For example the back of the capsule of the elbow joint is loose when the forearm is in a straight line with the upper arm and becomes taut when the forearm is bent. The capsule and ligaments on the sides of the elbow joint are taut in all positions of the forearm.

The synovial membrane produces a fluid called *synovia* (*synovial fluid*) which is a lubricant and prevents friction between the bone ends. It also nourishes the articular cartilage which has no blood supply of its own.

The cartilage is very smooth and consequently reduces friction. The cartilage is also deformable. Pressure between the joint surfaces results in a change in the shape of the articular cartilage which resumes its normal shape when pressure is removed.

Intracapsular discs, if present, may make incongruous surfaces more congruous as in the knee and mandibular joints. They are sometimes found where movement takes place in more than one plane as in the mandibular and wrist joints. They are also regarded as *spreaders* of the synovial fluid in the joint so that there is always a film of lubricant between the joint surfaces.

Bursae are found outside the joint where pressure is likely to occur between ligaments and the bone, or between ligaments and tendons of muscles attached near the joint. The cavity of a bursa may be continuous with the cavity of the joint which means that the synovial lining of the bursa is continuous with that of the joint.

Classification of synovial joints. Synovial joints are classified according to the shape of the joint surfaces or the movements which occur at the joint. A *ball and socket joint* (Fig.32a,d) is one in which part of a sphere fits into a concave socket which is spherical in shape. The shoulder and hip joints are of this type in which movements about all three axes (transverse, anteroposterior and longitudinal) can occur.

An *ellipsoid joint* (Fig.30d,e) is one in which one surface is concave in two planes and the other convex in two planes and the general shape of these surfaces is elliptical. As a result movement can take place about two axes, transverse and anteroposterior, but not about a longitudinal axis. The wrist joint is of this type.

A *condyloid joint* is one in which two convex, oval surfaces fit into two concave, oval surfaces. This type of joint is seen in the knee joint and interphalangeal joints of the fingers and toes (Fig.32d,f).

A *saddle-shaped joint* is one in which one joint surface is concave in one direction and convex in another and the other joint surface is shaped in the opposite directions. The surfaces may be described as reciprocally concavoconvex (Fig.36b). The carpometacarpal joint of the thumb is saddle-shaped and movements about all three axes are possible. Those about a longitudinal axis are much more limited than in a ball and socket joint.

A *hinge joint* (Fig.30c,d) is one in which movement occurs about a transverse axis. The elbow and ankle joints are of this type.

A *pivot joint* is one in which movement about a longitudinal axis occurs. The joint between the odontoid process of the axis and anterior arch of the atlas (completed by a ligament passing behind the odontoid process) (Fig.36a), and the superior and inferior radio-

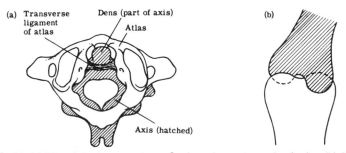

Fig.36. (a) Pivot joint between dens of axis and anterior arch of atlas, (b) Saddle-shaped joint.

ulnar joints at which the radius rotates round the ulna are joints of this type.

A *gliding joint* (Fig.32f) is one in which the joint surfaces are almost flat and movements may take place in all directions, but only to a very limited extent. At some joints of this type almost no movement occurs, for example, at joints between several of the tarsal bones.

MOVEMENTS AT JOINTS. From the anatomical position forward movements about a transverse axis are called *flexion* and backward movements are called *extension*. Forward bending of the head on the neck, the neck on the trunk and the trunk on the hips are called flexion, as are forward raising of the upper arm at the shoulder joint, forward bending of the forearm at the elbow, of the hand at the wrist and the fingers into the palm. Similarly forward raising of the lower limb at the hip is flexion. A change takes place lower down the lower limb. Backward bending of the leg at the knee, downward movement of the foot at the ankle and bending downwards of the toes are also flexion movements although the movements of the foot and toes are better referred to as *plantar flexion*. All the opposite movements are called extension but a better term for those of the foot and toes is *dorsiflexion* (Fig.37a,b).

Abduction is a movement away from the midline about an antero-posterior axis and *adduction* is a movement towards the midline about the same axis. The upper limb can be abducted through 180° from the side of the body to a position alongside the head. There is no abduction at the elbow and the hand can be abducted at the wrist to a limited extent. Abduction of the lower limb at the hip to about 60° is possible but there is no abduction at the knee and ankle.

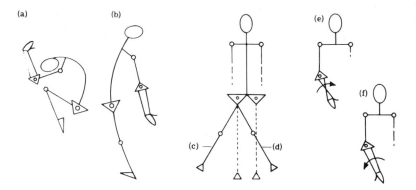

Fig.37. (a) Flexion, (b) Extension, (c) Abduction, (d) Adduction, (e) Pronation, (f) Supination.

Adduction of the upper limb is restoration of the abducted limb or bringing the upper limb across the front or back of the chest. Adduction of the hand at the wrist is much more extensive than abduction. The lower limb can be adducted at the hip by bringing it across the front of the other limb (Fig.37c,d).

Rotatory movements take place about a longitudinal axis and are possible at the shoulder and hip joints because they are ball and socket joints. In *lateral rotation* the anterior surface of the limb is turned laterally and in *medial rotation* it is turned medially. In the forearm the radius can rotate medially and cross over the ulna. This is called *pronation*. The opposite movement which brings the radius back to its position parallel and lateral to the ulna is called *supination* (Fig.37e). The head can rotate together with the atlas on the axis to the right and to the left. The trunk can be rotated to the right and to the left because each vertebra can rotate a little on the neighbouring vertebra.

Stability at joints depends on the shape of the joint surfaces (for example the hip joint), the ligaments (for example the hip and the knee joints), and the muscles (for example the shoulder joint).

The structure and function of striated muscles

The structure of striated muscle has already been considered. Further points to be noted are that muscles are attached to bones by means of connective tissue and that individual fibres of the muscle are attached to a connective tissue framework within the muscle itself. Although the muscles are attached by connective tissue, the term *tendon* is only used to describe the attachment if a visible structure is seen. Tendons are usually rounded and thick. The tendon of the biceps muscle can be seen and felt from in front of the elbow if the forearm is bent against resistance. A very flat tendon associated with a muscle in the form of a sheet is called an *aponeurosis*. This is seen in the muscles of the anterior abdominal wall. The *origin* of a muscle refers to the stationary part of the muscle and is usually proximal and the *insertion* which is usually distal refers to the moving part. The word *attachment* for both ends is better, since frequently the origin moves and the insertion remains stationary.

Muscle fibres can contract to about half their length. A muscle, half of which is tendon, can therefore contract to three-quarters of its length (Fig.38a,b). Another factor in determining the amount of shortening of muscle is the way in which the muscle fibres are arranged. If the fibres are in the long axis of the muscle, shortening is maximal (Fig.38c). If, however, there is a large number of short fibres attached obliquely into the side or sides of a tendon running the length of the muscle the power of the muscle is increased and its shortening decreased. These are called *pennate muscles* because they resemble a feather (Fig.38d,e).

When muscles contract they act on the bones, which may be compared with levers. The bones move at the joints acting as fulcra (Fig.39a,b,c). The further away the insertion of the muscle from the

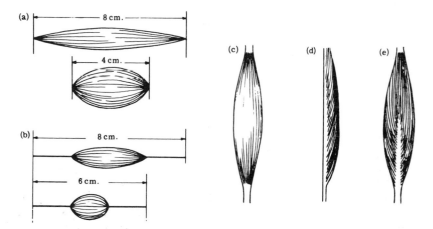

Fig.38. (a) Shortening of muscle to half its length, (b) Limitation of shortening if muscle has tendons, (c) Strap muscle, (d) Unipennate muscle, (e) Bipennate muscle.

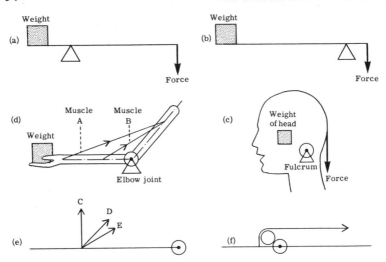

Fig.39. Less force is required to raise the weight in (a) than in (b), (c) Weight of head is raised as it pivots at fulcrum (atlanto-occipital joints) by a force (muscles attached to back of head), (d) Muscle A can exert a greater lift than muscle B (e) C exerts a greater vertical pull than D which is greater than E, (f) A horizontal pull is converted into a vertical pull by a pulley.

fulcrum the more powerful is the effect of contracting the muscle and the less is the range and speed of movement (Fig.39a). The nearer the insertion of the muscle to the fulcrum the less the power of movement and the greater is the range and speed of the movement. Another factor is the line of pull of the muscle. The most efficient line of pull is at right angles to the lever. Obviously muscles must pull obliquely in the body but by means of a pulley-like arrangement the line of pull is altered to one which is much nearer a right angle (Fig. 39e,f).

THE GROUP ACTION OF MUSCLES. When a movement is performed, the muscles actually performing the movement are called the *prime movers*. In almost all movements some muscle or muscles have to lengthen or relax in order to allow the prime movers to act. These are called the *antagonists*. For example, the forearm is bent at the elbow by means of a muscle spanning the front of the elbow. This muscle acts as a prime mover but the muscle spanning the back of the elbow has to lengthen in order to permit this movement to take place. Frequently a muscle acts against gravity which may then be regarded as the prime mover and the movement is controlled by the antagonists. If the upper limb is raised sideways to a right angle the movement is produced by the muscle on the outer side of the shoulder. (This is called *concentric contraction*.) If, however, the arm is then lowered to the side again, it is the same muscle which controls the movement by gradual relaxation. (This is called *eccentric contraction*.) *Synergists* are muscles which contract in order to prevent an unwanted movement at another joint by the prime movers. Many muscles act on more than one joint and their contraction would produce an unwanted movement unless the second joint was fixed. For example the muscles which bend the fingers, cross the front of the wrist joint. When grasping firmly with the fingers, the muscles on the *back* of the wrist prevent the hand from bending forwards. Such muscles are also called *fixators* although this term is better used for muscles which fix a joint not acted on by the prime movers.

THE PHYSIOLOGY OF MUSCLE CONTRACTION. In the living body striated muscle requires an intact motor nerve supply in order to contract. The motor neurones have their cell body in the *anterior horn* of grey matter in the spinal cord and their fibres leave the spinal cord in the anterior roots to join the spinal nerves and be distributed to the muscle fibres. A *motor unit* consists of an anterior horn cell, its axon and all the muscle fibres supplied by the branches of the axon. Motor units vary in size and may contain as few as five, and as many as 2,000, muscle fibres. The terminal branches of the motor nerve fibre make contact with a muscle fibre at the motor end plate. An impulse travels along the nerve fibre and when it reaches the motor end plate *acetyl choline* is released. This results in the spread of the impulse along the muscle fibre which then contracts. The spread of the impulse and subsequent recovery result in changes in the electric potential of the muscle fibre and these electric changes can be picked up, amplified and recorded. This is known as *electromyography*. The electric changes occur before the fibre contracts. By means of electromyography the action of muscles can be studied.

If a motor nerve is stimulated once, the muscle supplied by the nerve contracts once. This involves a short latent period (0·012 sec) followed by the contraction and relaxation of the muscle. This lasts 0·1 sec equally divided between contraction and relaxation. If a second stimulus is applied before the muscle has fully relaxed summation results so that the contraction is increased and prolonged. If the stimuli reach 50/sec then a smooth sustained contraction results. This is referred to as *tetanus*. Maximum contraction of a muscle is achieved by contracting all the motor units at a maximum rate of frequency. The contraction is sustained because all the motor units do not contract at one time, that is, they contract asynchronously. Eventually fatigue will result because of lack of oxygen and the accumulation of the products of metabolism. This occurs fairly quickly in a sustained contraction because the blood flow in the muscle is cut off with the result that oxygen cannot reach the muscle and the metabolites cannot be removed. If the muscle is contracted and relaxed alternately fatigue will eventually occur but much more slowly.

The contraction is said to be *isotonic* if the muscle shortens and *isometric* if it remains the same length but increases its tension. An example of isotonic contraction is seen when the forearm is bent rapidly at the elbow and the muscles producing the movement shorten. If the forearm is held bent at the elbow and a weight is put into the hand but the forearm is not allowed to move, the additional contraction of the muscles preventing the straightening of the forearm is an isometric contraction, that is, the muscle contracts more strongly but there is no shortening of the muscle.

It is important to realize that when muscles contract heat is produced. The contraction of muscles therefore contributes to the maintenance of the temperature of the body. The heat is produced during the activation, contraction and relaxation of the muscle and after the mechanical events are completed. The amount of energy lost as heat is about 75 per cent of the original energy available. Chemical processes which occur in the muscle result in restoration of about 50 per cent of the energy liberated as work and heat. Shivering is usually an involuntary contraction of striated muscle in order to produce heat which the body requires.

THE NERVE SUPPLY OF MUSCLE. When considering the nerve supply of muscles it must be appreciated that striated muscle has an afferent (sensory) as well as an efferent (motor) nerve supply. The motor nerve supply has already been considered and involves the anterior horn cell (or its equivalent in the brain) and its terminations at the motor end plates of the muscle fibres supplied by it. The sensory nerve supply involves the muscle spindles found throughout the muscle. Penetrating the fibrous capsule of a muscle spindle and winding round the middle of the spindle is the large peripheral process of a sensory neurone, with its cell body in a posterior root ganglion and its central process going to the spinal cord in a posterior nerve root. Stretching of the whole muscle or contraction of the muscle fibres in the spindle sends an impulse along this sensory neurone to the spinal cord. This neurone may link up directly with the motor neurone supplying the muscle or link up with other neurones so that the impulse may go to many parts of the central nervous system. The direct link with the motor neurone supplying the muscle produces a *reflex contraction* of the muscle itself. By this means information about the extent to which muscles are being stretched is conveyed to the nervous system. There are also branched sensory nerve endings lying amongst the muscle fibres near the attachment of the muscle to its tendon. These also convey information to the central nervous system about the state of the muscle either when it is contracted or stretched.

MUSCLE TONE. *Muscle tone* is a term used to describe the slight state of contraction which is found in muscles even at rest. It is said to be due to the contraction of a small number of motor units. On the assumption that the same motor units would become fatigued if continuously contracting, it is suggested that tone is due to the contraction of alternating motor units. Tone in muscles may render the muscles able to deal with greater demands and is often said to place the muscles in a state of readiness. Tone disappears if the nerve supplying the muscle is cut. It is said to be increased or decreased in many conditions affecting the central nervous system. On the other hand there is very little evidence that there are contracting motor units in muscles at rest, and muscle tone is probably better defined as the response of a muscle to stretch.

4

The locomotor system. II

The joints and muscles of the skull

THE TEMPOROMANDIBULAR JOINT. The separate bones of the skull are joined to each other by sutures and no movements take place between these bones. The exception to this is the joint between the mandible and inferior surface of the temporal bone, the *temporomandibular joint*, at which movements such as opening and closing the mouth can take place. This is a synovial joint. The capsule is attached to the edges of the articular areas above and below, that is, to the edge of the mandibular fossa and to the articular tubercle above, and to the *neck* of the mandible below (the part of the condyloid process below the head). The lateral part of the capsule is thickened and forms a ligament (intrinsic) and there is an important extrinsic ligament passing from the base of the skull to near the mandibular foramen on the inside of the ramus of the jaw.

This joint also has an articular disc of fibrous tissue. Its upper surface is convex behind and concave in front where it fits into the mandibular fossa and on to the articular tubercle respectively, and concave below where it fits over the head of the mandible. It is attached peripherally to the capsule so that it divides the joint cavity into two. It moves with the head of the mandible when it is pulled forwards.

The muscles of mastication move the lower jaw at the temporo-mandibular joints. The *masseter muscle*, passing from the zygomatic arch to the outer surface of the ramus of the jaw, can be felt on the side of the face on clenching the teeth (Fig.40). The *temporalis muscle*, which passes from the lateral aspect of the skull over the temporal bone downwards to the coronoid process of the mandible, can be felt above and behind the eyebrow on clenching the teeth

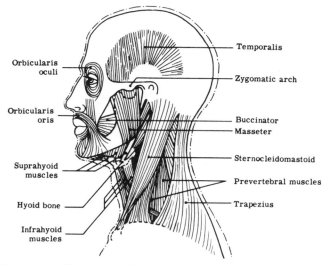

Orbicularis oculi

Orbicularis oris

Suprahyoid muscles

Hyoid bone

Infrahyoid muscles

Temporalis

Zygomatic arch

Buccinator
Masseter

Sternocleidomastoid

Prevertebral muscles

Trapezius

Fig.40. Muscles of head and neck.

(Fig.40). Deep to both these muscles are two *pterygoid muscles* (Fig.41). The floor of the mouth is formed mainly by the two *mylohyoid muscles* which are attached laterally to the inner side of the body of the mandible and pass towards the midline where they meet in a raphe (this word is used to describe the intermingling and crossing of fibres from each side). More posteriorly the muscle is attached to the hyoid bone. All these muscles are supplied by the mandibular division of the trigeminal nerve, the fifth cranial nerve.

The movements of the jaw may be summarized in the following way:

a. opening and closing the mouth;

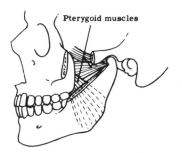

Pterygoid muscles

Fig.41. Pterygoid muscles.

b. pulling the jaw forwards (protraction) and backwards (retraction);

c. side-to-side movements.

Opening and closing the mouth takes place about a transverse axis passing through the middle of the rami. This corresponds approximately to the position of the mandibular foramina. In opening the mouth the head together with the disc is pulled forwards by one of the pterygoid muscles and gravity assists this movement. If a finger is placed in front of the ear-hole the head can be felt hitting the finger when the mouth is opened. Against resistance, opening of the mouth is aided by the mylohyoid muscles. Closure of the mouth is carried out by the masseter, temporalis and the second of the pterygoid muscles. This is a very powerful movement. The lower jaw is pulled forwards by two of the pterygoid muscles and pulled back by the horizontal back fibres of the temporalis muscles. When the jaw is pulled to the left it rotates about a vertical axis through the left condyle and the movement is carried out by the right pterygoid muscles. Movement to the right is produced by the left pterygoid muscles.

These jaw movements occur in biting and chewing. The jaws are held together during swallowing and are moved to some extent during speech.

The mandible is relatively easily dislocated. This usually occurs when the mouth is opened widely, for example, when yawning. The dislocation is on one side as a rule and consists of the head moving forwards and becoming fixed in front of the articular tubercle. If the mandible is forced downwards by means of the thumb in the mouth the dislocation is usually easily reduced. Care should be taken to remove the thumb from the patient's mouth before the lower jaw snaps shut.

THE MUSCLES OF FACIAL EXPRESSION. The muscles of facial expression are found under the skin of the face, the scalp and the front of the neck. They may also be said to be related to the orifices of the skull, namely the mouth, nasal cavity, orbit and external ear (Fig.40). There are many muscles related to the lips. The *orbicularis oris muscle* surrounds the mouth but consists of a number of muscles entering the upper and lower lips and the angles of the mouth as well as circular muscle fibres within the lips themselves. Muscles enter the upper lip from above and the lower lip from below. More muscles enter the angle of the mouth horizontally and pass into both lips. All these muscles are responsible for the many lip movements associated with speech and facial expression. The *buccinator*, a deep muscle at the side of the mouth in the cheek, also pushes the food between the teeth. If it is paralysed, food collects between the cheek and the gums.

The *orbicularis oculi muscle* is found in the eyelids and round the outside of the orbit. Its fibres run round the orbit and eyelids and are attached to bone only on the medial side. The fibres in the eyelids are responsible for closing the eyes gently and those round the orbit for screwing up the eyelids tightly. The muscles of the nose, external ear and scalp need not be considered. It may be pointed out that the nose and scalp can be moved by everybody but the auricle only by a limited number of people.

All the muscles of facial expression are supplied by the *facial nerve*, the seventh cranial nerve. In *Bell's palsy* this nerve is affected and the muscles of one side of the face are paralysed. The result is that the face is pulled over to the unaffected side because the normal symmetry of the face depends on the balanced pull of the muscles of the two sides. The patient cannot, among other things, wrinkle his forehead, frown, close his eyes, twitch his nose, smile or whistle with the affected side of his face, and as has already been pointed out food collects between the gums and cheek.

The muscles of the neck

In the front of the neck, the muscles are conveniently grouped into those above the hyoid bone (*suprahyoid*) and those below (*infrahyoid*). The infrahyoid muscles tend to fix the hyoid bone so that the suprahyoid muscles can contract and move the tongue and jaw. These muscles are also used in very complicated ways during swallowing and speaking. Some of these muscles are seen in Fig.40. There is also a group of *prevertebral muscles* (Figs.40,42). Some are attached to the front of the cervical vertebrae and are used in movements of the vertebral column in this region, and some pass from the cervical vertebrae to the first and second ribs and are involved in movements of these ribs in respiration.

A very important muscle is seen on the side of the neck passing from the back of the head behind the ear to the inner part of the clavicle in front. It is called the *sternocleidomastoid muscle* (Figs.40,42). The right muscle turns the head to the left side and tilts the chin upwards. It is an abnormal contraction or shortening of this

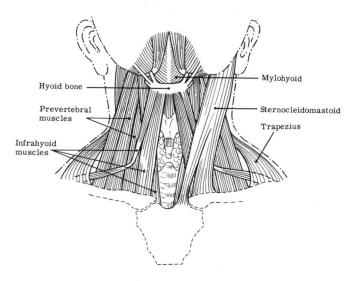

Fig.42. Muscles of front of neck.

muscle which produces the typical deformity of *wry neck* (*torticollis*).
Both muscles acting together bend the head and neck downwards
(flexion).

At the back of the neck there are several layers of muscles. The
superficial muscles, in two layers, are associated with the pectoral
girdle and the upper limb. The main one, which forms the line of the
neck between the head and the shoulder, is the *trapezius muscle*
(Figs.40,49). It is attached medially to the midline of the back, from
the head to the lowest thoracic vertebra, and laterally to the scapula
and clavicle. It pulls back and elevates the shoulder. The deeper
layer of muscles elevates the scapula.

Deep to these two layers are muscles which act on the head and the
vertebral column, the movements of which will be dealt with as a
whole. It should be mentioned that there are many small muscles
between the atlas and axis and occipital bone which lie in front
and behind in the deepest parts of the neck. They are responsible for
fine movements between the head and atlas, and the atlas and axis.

The joints and muscles of the thoracic cage related to respiration

THE JOINTS. The ribs articulate posteriorly with the vertebrae and
anteriorly with a costal cartilage which in turn articulates with

either the sternum (cartilages 1–7) or a costal cartilage above (cartilages 8–10) or end in the abdominal wall (cartilages 11–12) (Fig.29).

Posteriorly there are two joints. The head of the rib articulates with the facets on the adjacent edges of the bodies of the thoracic vertebrae and with the intervertebral disc between them (Fig.43a). This joint is synovial of the plane or gliding type. They therefore have a capsule lined by synovial membrane and the articular surfaces are covered by articular cartilage. The tubercle of the rib articulates with the transverse process of the appropriate thoracic vertebra (Fig.43a). This is also a synovial, plane joint except for the eleventh and twelfth which are fibrous joints.

The first rib is joined to the manubrium by its cartilage and constitutes the only primary cartilaginous joint in the adult. The costal cartilages articulate with the sternum by synovial joints.

The manubriosternal joint has already been described. It is a secondary cartilaginous joint (see p. 47).

THE RESPIRATORY MUSCLES. The respiratory muscles are mainly the *intercostal* and the *diaphragm*. Other muscles are also involved in respiration, for example, the muscles of the abdominal wall and, in forced breathing, almost any muscle capable of acting on the thoracic cage.

The intercostal muscles. The intercostal muscles are in three layers and lie between the ribs (Fig.43b). The *external* passes downwards and forwards between the lower border of one rib and the upper border of the rib below in the spaces between the ribs. There are eleven pairs of these muscles. An *internal intercostal muscle* is deep to an external and has similar attachments. Its fibres run at

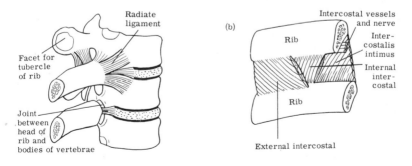

Fig.43. (a) Articulations of posterior end of rib with vertebrae, (b) Intercostal muscles, vessels and nerve.

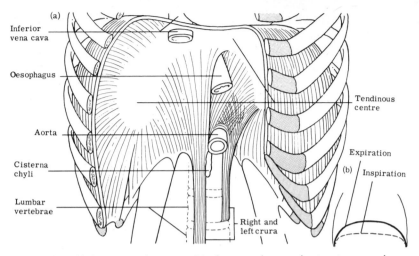

Fig.44. (a) Diaphragm partly removed in front to show main structures passing through it, (b) Downward movement of diaphragm in inspiration, upward movement in expiration.

right angles to those of the external. There are eleven pairs of these. The innermost intercostal muscle is called the *intercostalis intimus*. These are often absent in the upper spaces and are not so extensive horizontally. Their fibres run parallel to those of the internal intercostal muscles.

The Diaphragm. The diaphragm is a dome-shaped muscle with a tendinous centre and it separates the abdominal from the thoracic cavity (Fig.44a). The muscle fibres are peripheral and are attached to the circumference of the thoracic outlet. Anteriorly they are attached in the midline to the deep surface of the xiphoid process (*sternal part*). They are also attached to the deep surface of the lower six costal cartilages and ribs anteriorly, laterally and posteriorly (*costal part*). The most posterior fibres of the diaphragm are attached to ligaments on the posterior abdominal wall (*lumbar part*). Longitudinal muscle fibres are also attached to both sides of the bodies of the lumbar vertebrae. These are called the *crura* (singular—*crus*). The right crus is attached to the upper three lumbar vertebrae, the left to the upper two. All the muscle fibres pass upwards from their attachments and converge on the central tendon to which they are attached. The more medial fibres of the crura cross. The central tendon is trefoil-shaped with an anterior and two lateral leaves.

A large number of structures pass between the abdomen and thorax through or outside the diaphragm (Fig.44a). In the midline posteriorly at the level of the twelfth thoracic vertebra the *thoracic aorta* passes downwards behind the diaphragm to become the *abdominal aorta* and through the same opening the *thoracic duct* (a large lymphatic vessel) passes upwards. The *oesophagus* passes downwards through the muscular part of the diaphragm about 3 cm to the left of the midline at the level of the tenth thoracic vertebra. The *inferior vena cava* passes upwards through the right leaf of the central tendon at the level of the eighth thoracic vertebra. The upper surface of the diaphragm is related to the right and left *pleurae* and *lungs* on either side and in the middle to the *pericardium* and *heart*. There is a firm attachment between the pericardium and the central tendon. Inferiorly the right half of the diaphragm is related to the right lobe of the *liver* and behind that to the right *kidney*. The left half is related to the upper part of the *stomach*, the *spleen* and the left *kidney*.

THE MECHANICS OF RESPIRATION. The term *mechanics of respiration* refers to the movements of the walls of the thorax and the diaphragm during respiration and involves consideration of most of the muscles, bones and joints of the thorax. The act of *respiration* in a healthy adult under normal conditions takes place about 18 times per minute and consists of *inspiration* in which air is taken into the lungs, and *expiration* in which air is expelled from the lungs. Inspiration is due to the contraction of the muscles, especially the diaphragm, so that the capacity of the thorax is increased and the lungs expand. In quiet breathing the total diaphragmatic excursion is about 2 cm and in deep breathing about 9 cm. The downward movement is due to the peripheral muscular part pulling the central tendinous part (Fig.44b). The result is that the vertical diameter of the thorax is increased. The costal part of the diaphragm may also move the lower ribs upwards and outwards thus increasing the transverse diameter of the thorax. The descent of the diaphragm involves the descent of the abdominal viscera and the relaxation of the muscles of the anterior abdominal wall. The diaphragm is the main muscle of quiet respiration.

What exactly the intercostal muscles do in quiet respiration it is difficult to determine. The lower ribs may move outwards and the upper ribs may be elevated by some of the deep neck muscles attached to the first and second ribs. The movements, which are slight or absent in quiet respiration, increase the size of the thoracic cavity but there is little evidence that the intercostal muscles are responsible for these movements.

Expiration is largely a passive act. The muscles at the end of inspiration relax and the diaphragm rises into the thorax. The ribs are lowered, if they have moved, due to relaxation of the muscles which moved them. The stretched lungs, which contain a large number of elastic fibres, become smaller. The abdominal muscles sometimes contract at the end of expiration and push the abdominal viscera upwards towards the thorax.

In deep inspiration the diaphragm moves down very much more than in quiet inspiration, the upper ribs move in such a way that the anteroposterior and transverse diameters of the thoracic cage are increased and the lower ribs move so that the transverse diameter is increased (Fig.45a,b). It is the neck muscles referred to above

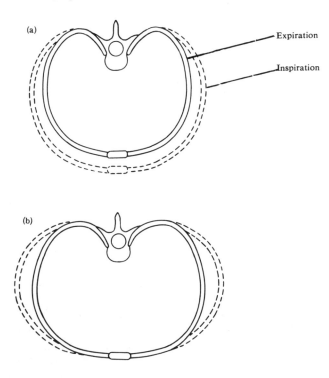

Fig.45. (a) Movement of upper part of thoracic wall in inspiration showing increase in transverse and anteroposterior diameters, (b) Movement of lower part of thoracic wall in inspiration showing increase in only transverse diameter.

which move the ribs assisted by the sternocleidomastoid muscles. The intercostal muscles also contract in both inspiration and expiration and may prevent the sucking in and blowing out of the intercostal spaces on inspiration and expiration respectively. In deep expiration the diaphragm moves upwards much more than in quiet expiration. This is due to a marked contraction of the muscles of the anterior abdominal wall. The *pectoralis major, pectoralis minor* and *serratus anterior muscles,* which are attached to the humerus or scapula at one end and to the upper ribs at the other, may help to increase the movement of the ribs.

In forced respiration, that is, where there is obstruction to breathing, the contraction of all the muscles mentioned is exaggerated in order to overcome the obstruction.

It should be added that the movements in respiration are assisted by pressure differences between the atmosphere and the air in the lungs, and by the negative pressure in the potential space between the lungs and the chest wall.

THE NERVE SUPPLY OF THE MUSCLES OF RESPIRATION. One branch of the cervical plexus is called the *phrenic nerve.* Its fibres come from the anterior primary rami of the third, fourth and fifth (mainly fourth) cervical nerves and the nerve passes downwards through the neck and thorax to the diaphragm. It takes this peculiar course because the diaphragm originally developed in the region of the neck and was pushed downwards by the developing heart and lungs. When the position of a muscle changes it invariably takes its nerve supply with it. The anterior primary rami of the thoracic spinal nerves form the *intercostal nerves* which run round the chest wall in an intercostal space under cover of the lower part of the rib and between the internal intercostal muscle and the intercostalis intimus, together with an *intercostal artery* and an *intercostal vein* (Fig.43b). Each intercostal nerve supplies the intercostal muscles of its own space. The lower five intercostal nerves and the anterior primary ramus of the twelfth thoracic nerve (*subcostal nerve*) pass into the anterior abdominal wall and supply its muscles. The intercostal nerves also supply the skin of the thorax and abdomen. The intercostal arteries come from the thoracic aorta except the first two which come from the *subclavian artery.* The intercostal veins pass backwards towards the vertebral column and form larger longitudinal vessels which end in the *superior vena cava.*

The muscles of the trunk and the movements of the vertebral column

There are several muscles on the front of the chest and on the back of the trunk which are associated with the pectoral girdle and upper limb. These will be considered later. The muscles mainly concerned with movements of the vertebral column are the anterior abdominal muscles and the muscle mass lying on either side of the vertebral column extending from the sacrum to the occipital bone

THE MUSCLES OF THE ANTERIOR ABDOMINAL WALL. The muscles of the anterior abdominal wall fall into two groups, lateral muscles consisting of muscles, corresponding to the three layers of intercostal muscles, and medial muscles one on either side of the midline (*rectus abdominis muscles*). The lateral group lie posteriorly, laterally and anteriorly. They are in three layers and form flattened sheets which are aponeurotic in front. The three muscles of the lateral part of the abdominal wall split to enclose the rectus abdominis muscle and thus form the *rectus sheath* (Fig.46). They meet in the midline to form the *linea alba*. The outer of the three muscles is called the *external oblique*, the middle the *internal oblique* and the inner the *transversus abdominis*. These muscles are attached posteriorly to the lower part of the vertebral column, below and in front to the pelvis, and above and in front to the ribs and costal

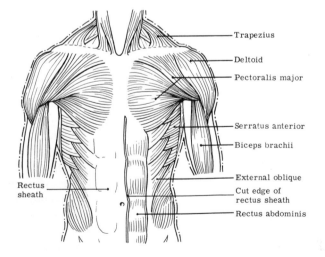

Fig.46. Muscles of front of trunk.

cartilages. The transversus abdominis muscle lies *within* the ribs, the other two outside. Their functions are to produce movements of the vertebral column, to increase intra-abdominal pressure and thus help in emptying hollow viscera of their contents, to retain the position of the viscera, and to assist in respiration. They are supplied by the lower intercostal nerves.

The rectus abdominis muscle is the medial muscle. It is long and strap-like, extending from the ribs to the pubic part of the pelvis. If one lies on one's back and tries to raise one's lower limbs or shoulders from the couch these muscles will be found to contract very strongly. Abdominal incisions are frequently made vertically through the linea alba or to the side of the midline through the rectus sheath. In the latter case the muscle is pulled to one side or split and the incision is then continued through the posterior part of the rectus sheath.

THE POSTERIOR VERTEBRAL MUSCLES. The posterior vertebral muscles are complicated. Collectively they are known as the *erector spinae* (*sacrospinalis*) *muscle*. They are in several layers and in several more or less parallel columns but their details need not be considered. Some muscle bundles are vertical and some are oblique. They pass between

    a. the pelvis and the ribs,

    b. the ribs and the vertebrae,

    c. the different parts of the vertebrae,

    d. the vertebrae and the skull.

THE MOVEMENTS OF THE VERTEBRAL COLUMN. The joints between the bodies of the vertebrae have already been considered. They are secondary cartilaginous joints. There are synovial joints of the gliding type between the articular processes of the vertebrae. The combination of cartilaginous and gliding synovial joints means that movements between adjacent vertebrae are very limited. The joints between the head and atlas and between the atlas and axis are special and permit quite extensive bending forwards and backwards and rotation respectively. It should be remembered that even if only 4° of bending forwards (flexion) are permitted between each vertebra from the axis (second cervical) to the sacrum, there will be a total of 96° of flexion between these two bones ($24 \times 4°$). The movements possible in the vertebral column are flexion and extension,

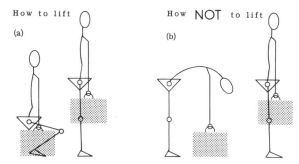

Fig.47. (a) Lifting by means of straightening the bent lower limbs, (b) Lifting by means of straightening the flexed trunk.

rotation to the right and left, and lateral bending to the right and left. In the upright position bending forwards is controlled by the back muscles acting against gravity. When lying on one's back bending forwards is brought about by the anterior abdominal muscles.

That movement between adjacent vertebrae is limited is emphasized by the ease with which an intervertebral disc can be damaged. Sudden bending forwards can cause this condition. People are advised to lift heavy objects by straightening the segments of the lower limbs and not by straightening the back (Fig.47a,b). When this advice is given, it is often forgotten however, that by bending forwards the reach of the upper limb is increased and this may be so important that the risk of damage to the intervertebral discs is overlooked.

THE INGUINAL CANAL. Before leaving the anterior abdominal wall some mention must be made of the lowest part of the external oblique muscle. It is attached to the ilium laterally and to the pubis medially. Between these two bony attachments it has a lower free edge called the *inguinal ligament*. There is a space between this ligament and the pelvis and through this space many structures pass into and out of the abdominal cavity (Fig.48). Above the medial part of the ligament there is a canal passing through the muscles of the abdominal wall, the *inguinal canal*. In the male just before birth the testis which develops in the abdominal cavity passes down this canal into the scrotum and takes with it its duct and blood supply. The peritoneum which lines the abdominal cavity has a process passing down behind the testis into the scrotum (Fig.49a,b). Normally the end of this process surrounds the testis and the rest of the peri-

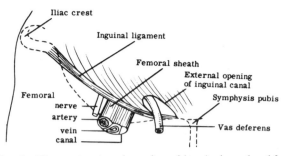

Fig.48. Inguinal ligament, external opening of inguinal canal and femoral sheath.

toneum up to the hole in the abdominal wall disappears. If it persists it may later form the *sac* of a *hernia* (a *hernial sac* is the protrusion of the lining membrane of any cavity of the body). This is the commonest form of hernia, a *congenital inguinal hernia*. In the female there is also an inguinal canal and a process of peritoneum but it is much smaller than in the male. Even in the female, a congenital inguinal hernia is the commonest form of hernia.

The joints and muscles of the upper limb

THE PECTORAL GIRDLE AND SHOULDER JOINT. On the front of the chest a large muscle, the *pectoralis major*, passes from the clavicle and sternum to the upper end of the humerus (Figs.46,50b).

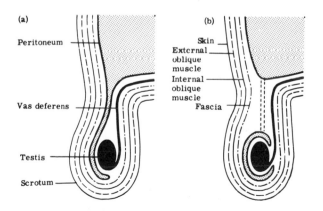

Fig.49. (a) Layers of abdominal wall passing down into scrotum together with testis and peritoneum, (b) The disappearance of peritoneum between inguinal canal and testis.

At the side of the shoulder joint the *deltoid muscle* is attached above to the clavicle and scapula and below to the humerus (Figs.46,50). On the back the trapezius muscle and a layer of muscles deep to it have already been mentioned (Fig.50). Another large muscle is attached to the vertebrae in the midline and covers the lower part of the back. It passes round the side of the trunk and upwards to the front of the humerus, and is called the *latissimus dorsi muscle* (Fig.50a). Many more muscles pass from the lateral edge and both surfaces of the scapula to the humerus. An important muscle, called the *serratus anterior*, passes from the sides of the upper ribs backwards between the chest wall and the scapula and is attached to the inner border of the scapula (Figs.46,50b).

The movements at the ball and socket shoulder joint between the head of the humerus and the glenoid cavity of the scapula usually involve movements of the scapula on the chest wall which involves

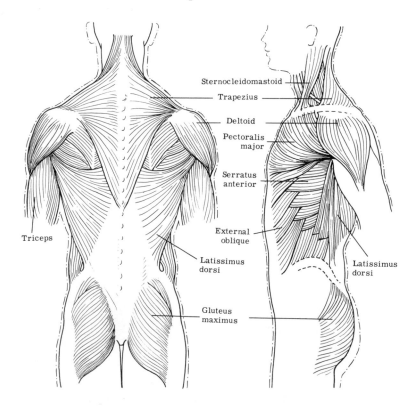

Fig.50. (a) Superficial muscles of back, (b) Superficial muscles of neck and trunk.

movements of the joints between the clavicle and sternum (*sterno-clavicular joint*). At a ball and socket joint, flexion and extension, abduction and adduction, and rotatory movements about a longitudinal axis are all possible. One can raise the upper limb from a position alongside the trunk through 180° to a position alongside the head. In this movement the humerus moves on the scapula and the scapula moves on the chest wall. If the latter does not take place movement is limited to between 90° and 110°. The main muscles producing this movement are the deltoid moving the humerus on the scapula and the serratus anterior moving the scapula on the chest wall.

If the forearm is bent to a right angle the rotatory movements at the shoulder are not confused with the rotatory movements within the forearm. The shoulder rotatory movements are easily lost in many diseases of the shoulder region. Adequacy of the extent of the movements can be tested by asking the patient to scratch the middle of his back from below (medial rotation) and from above (lateral rotation).

The movements of the scapula on the chest wall without movement at the shoulder joint can be appreciated when shrugging the shoulders, bracing back the shoulders or thrusting forwards the outstretched upper limbs. The trapezius muscle pulls the shoulders upwards and also backwards and the serratus anterior pulls them forwards.

All the other muscles are important in producing the large variety of movements possible at the shoulder joint. Many of them play an important part in stabilizing this joint and are attached to the capsule of the joint as well as to bone. The head of the humerus is large compared to the glenoid cavity of the scapula and the humerus is relatively easily dislocated. This dislocation takes place in a downward direction, the one site where there are no muscles protecting the joint. There are muscles in front of, above and behind the shoulder joint. Downward dislocation may result in injury to the axillary (circumflex) nerve which passes backwards between muscles below the joint and supplies the deltoid muscle. Injury to this nerve produces an inability to abduct the humerus at the shoulder joint because the deltoid is paralysed.

THE MUSCLES OF THE UPPER ARM AND FOREARM; THE ELBOW RADIO-ULNAR AND WRIST JOINTS (Fig.51)

The elbow joint and its muscles. The most obvious muscle of the upper arm is the *biceps brachii* which can be seen bulging when the forearm is strongly flexed at the elbow. This muscle is attached above by two heads to the scapula (hence its name) and below to the upper end of the radius (Fig.46). Another muscle which bends the forearm is the *brachialis muscle* lying deep to the lower end of the biceps muscle and attached above to the humerus and below to the ulna (Fig.51a). These muscles bend the forearm but if necessary any muscle crossing the front of the elbow joint may be involved in a rapid or powerful movement. The *triceps muscle* (Fig. 50a) on the back of the upper arm has three heads, two large ones from the back of the humerus and one from the scapula. It passes downwards and is attached to the upper end of the ulna (Fig.51c). It straightens the bent elbow.

The elbow joint is a synovial hinge joint between the humerus above and the radius and ulna below. It is mainly the ulna which articulates with the humerus by a notch running on a pulley on the distal end of the humerus (Fig.30).

Pronation and supination of the forearm. There are three groups of muscles on the front of the forearm: those which pass between the radius and ulna, those which are attached above to the humerus and below to the metacarpal bones of the wrist and those attached above to the radius and ulna and going to the thumb and fingers. The first group is associated with a movement which pulls the radius round across the ulna and turns the hand so that it faces backwards (*pronation*) and the opposite movement which restores the radius to its position of being parallel to the ulna (*supination*). It should be noted that the biceps muscle is a powerful supinator. This muscle is said to be used to insert a corkscrew in a cork in a bottle (repeated supination of the forearm) and then to pull the cork out (flexion at the elbow joint). In pronation the round head of the upper end of the radius moves in a ring formed by a ligament and a notch at the upper end of the ulna (*superior radio-ulnar joint*) and the lower end of the radius moves round the circular lower end of the ulna.

The wrist joint and its muscles. The second group of muscles comes from the inner side of the lower end of the humerus and goes to the metacarpal bones. They flex the hand at the wrist, a movement which takes place partly at the joint between the radius and carpal bones and partly between the two rows of carpal bones.

On the back of the forearm there are two groups of muscles. One group is superficial and comes from the lateral side of the lower end

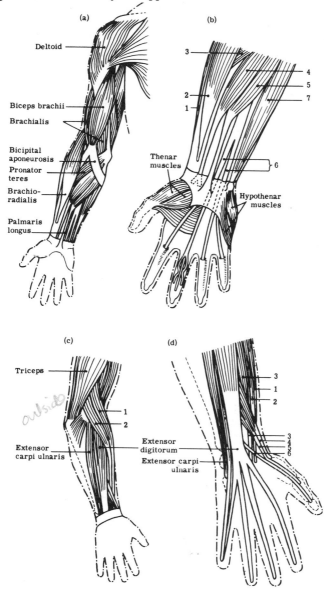

Fig.51. (a) Muscles on front of upper limb, (b) Some muscles of forearm and hand
(1) extensor carpi radialis longus (2) brachioradialis (3) pronator teres (4) flexor
carpi radialis (5) palmaris longus (6) flexor digitorum superficialis (sublimis)
(7) flexor carpi ulnaris, (c) Some muscles on back of upper limb, (1) brachioradialis (2) extensor carpi radialis longus, (d) Muscles and tendons on back of wrist
(1) brachioradialis (2) extensor carpi radialis longus (3) extensor carpi radialis
brevis (4) abductor pollicis longus (5) extensor pollicis brevis (6) extensor pollicis
longus.

of the humerus. Some of these muscles go to the back of the metacarpal bones and act on the hand at the wrist joint (Fig. 51c,d).

The wrist joint is an ellipsoid synovial joint between the radius
and the carpal bones. The movements which can take place at this
joint are bending forwards and backwards, and movement of the
hand towards the radial side (abduction) and towards the ulnar
side (adduction). The muscles which bend the hand forwards are
on the front of the forearm and those which bend it backwards are
on the back of the forearm. The same muscles produce abduction
and adduction but work in different groups. Abduction is produced
by the front and back muscles on the radial side and adduction
by the front and back muscles on the ulnar side. All these movements take place at the wrist joint and also between the two rows of
carpal bones.

The muscles and joints of the hand. The third group of anterior
muscles of the forearm goes to the phalanges of the fingers and bends
the phalanges into the palm of the hand (Fig. 51b). The muscle to
the thumb bends that digit across the palm of the hand.

Some of the superficial muscles of the back of the forearm go
to the back of the phalanges and straighten the bent fingers. The
deep group comes from the back of the radius and ulna and goes to
the thumb and index finger. They straighten these digits after they
are bent (Fig. 51d).

Many of the muscles which move the digits are long muscles and
come from the lower end of the humerus and the radius and ulna.
There are also many muscles in the palm of the hand itself, about
twenty altogether. There are three on the lateral side at the base of
the thumb (the *thenar eminence*) and three on the medial side (the
hypothenar eminence) (Fig.51b). There are twelve, four in each of
three layers, in the middle of the hand. Their attachments and actions
are complicated and they function in association with the long muscles
from the forearm. Some of these muscles pull the fingers apart and
bring them together again (the *interossei muscles* between the metacarpal bones).

There is one special movement in the hand which should be
mentioned. It is called *opposition* and consists of bringing the tip
of the thumb against the tip of any of the other fingers. In this
movement the thumb rotates about a longitudinal axis and is flexed
across the palm. The thumb nail looks forwards at the end of the
movement instead of facing laterally, its position at the beginning of

the movement. The digit to which the thumb is approximated is also bent. Holding things between the tips of the thumb and any of the fingers is associated with precise, delicate activities, whereas holding things firmly in the palm of the hand is associated with powerful movements. The thumb has special muscles for the movement of opposition.

The joints and muscles of the lower limb

THE JOINTS OF THE PELVIC GIRDLE. The pelvic girdle is firmly attached to the lower end of the vertebral column at the sacro-iliac joints (Fig.31). This is in striking contrast with the mobility of the pectoral girdle. The sacro-iliac joints are important in that the weight of the head, neck, upper limbs and trunk is transferred to the lower limbs through these joints. These joints are almost completely immobile and this immobility is associated with the stability required for maintaining the upright posture and for locomotion. However these joints can suffer slight displacement or tearing of their ligaments both of which can lead to considerable disability.

The two hip bones articulate in front at the symphysis pubis, a secondary cartilaginous joint which has already been described (p. 47).

THE HIP JOINT AND ITS MUSCLES. The hip joint is a ball and socket synovial joint between the round head of the femur and the deep socket formed by the acetabulum of the hip bone (Fig.32). The movements at this joint are flexion (forward movement in the sagittal plane), extension (the opposite movement), abduction (away from the body in the coronal plane), adduction (the opposite movement) and rotatory movements about a longitudinal axis.

There are muscles in front of, behind and on either side of the hip joint and several of them are attached to bones inside the abdominal cavity. One of these is the *iliopsoas muscle* which consists of the *iliacus muscle* from the ilium and the *psoas muscle* from the lumbar vertebrae (Fig.52a). It enters the thigh deep to the inguinal ligament and is attached to the femur. It is the main flexor of the thigh at the hip joint. The very large muscle of the buttock, the *gluteus maximus*, is attached medially to the back of the sacrum and laterally to the deep fascia of the thigh and also to the femur (Fig.52b). It is an extensor of the thigh. When standing upright, extension is limited to about 15°–20°. Swinging the free limb backwards at the hip is not a particularly important movement but standing up from

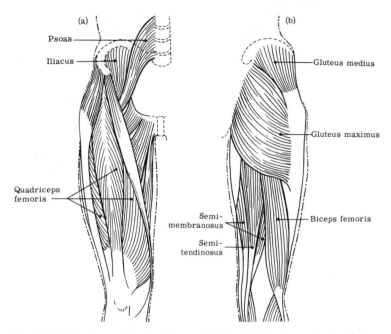

Fig.52. (a) Muscles on front of hip and thigh, (b) Muscles on back of hip and thigh.

the sitting position, or going up stairs, or walking up hill involves extension of the thigh from the flexed position. In all of these activities the body weight is being lifted upwards. The lower limb is also extended in each step in walking.

Abduction of the thigh at the hip is brought about by the *gluteus medius muscle* which is deep to the gluteus maximus muscle and is attached to the ilium and the side of the femur (Fig.52b). Again it can be said that lifting the free limb sideways is not a very important movement except perhaps in certain steps in ballet but the action of the gluteus medius muscle in preventing the unsupported side from falling when standing on one limb is of great significance. When standing on the right lower limb the right gluteus medius prevents the left side of the pelvis from falling. In many clinical conditions affecting the head and neck of the femur the action of this muscle is interfered with. The *adductor muscles* (Fig.52a) form a large mass on the inner part of the thigh and they are attached to the hip bone below and medially, and to the back of the femur.

The many small muscles deep to the gluteus maximus muscle on the back of the hip joint pass from the pelvis to the upper end of the femur and turn the lower limb outwards. The *gluteus minimus muscle* passing from the side of the pelvis to the front of the femur turns the limb inwards.

The hip joint is very stable due to the shape of the bones forming the joint, the very powerful ligaments strengthening the capsule and the many muscles round the joint. The socket for the head of the femur is deep and its edge embraces the head beyond its equator. In front of the joint there are three very powerful ligaments passing between the hip bone and the front of the femur. Dislocation of the head of the femur is a comparatively unusual occurrence.

THE MUSCLES OF THE THIGH; THE KNEE JOINT. The large muscle on the front of the thigh is called the *quadriceps femoris* because it has four different parts (Fig.52a). Three of these are attached to the femur and one to the hip bone. They cover the front and sides of the thigh and when followed downwards are seen to converge towards the patella (knee cap) which is a sesamoid bone in the tendon of the quadriceps femoris muscle. Beyond the patella, the tendon is renamed the *patellar ligament* which is attached to the upper end of the tibia. It is a very powerful muscle and it straightens the bent knee. On the back of the thigh are the *hamstring muscles* (Fig.52b). They are attached above to the ischial tuberosity of the pelvis and below they diverge. Two of them, on the medial side, go to the tibia and one, on the lateral side (the *biceps femoris muscle*) goes to the fibula. Their tendons can be seen and felt quite easily behind and above the knee when the knee is bent.

The knee joint is a synovial joint of the condyloid type. Actually the movements at this joint are more complicated than those found at other condyloid joints in the body, for example, the interphalangeal joints of the fingers. The neutral position of the thigh and the leg is when they are in a straight line. Backward movement of the leg on the thigh, bending the knee, is a movement of flexion and straightening the knee is extension. The change from the usual terminology is due to the fact that the lower limb is twisted as compared with the upper limb and the front of the knee, leg and foot correspond with the back of the elbow, forearm and hand. The great toe is medial, and the thumb, the comparable digit, is lateral. The muscles which flex the leg at the knee joint are the hamstring muscles which, as already stated, pass from the pelvis to the tibia and fibula. It should be noted that they span two joints, the hip and the knee. These muscles are also extensors of the trunk on the lower limb at the hip joint and they are used for this purpose before the gluteus maximus muscle is brought into action. Flexion at the knee joint is limited by contact of the back of the leg with the back of the thigh.

Extension of the leg on the thigh is brought about by contraction of the quadriceps femoris muscle. It is a very powerful muscle and its size is not related to straightening the leg on the thigh but to its control of the bending of the knee when sitting down and straightening the knee when standing up or going up stairs. Weakness of this muscle may result in the knee giving way when the weight of the body falls behind the axis of movement of the knee joint. This may occur in normal walking on the flat.

At the end of extension of the thigh on the leg or the leg on the thigh a rotatory movement takes place. This has often been referred to as *locking* of the knee in full extension. This rotation however, has no stabilizing effect as is well demonstrated by the childish trick of unexpectedly hitting the back of the knees of someone standing upright with full extension at the knee joints. Before flexion can take place, the leg has to be rotated to some extent on the thigh or vice versa, a movement which has been called *unlocking* of the knee. These rotatory movements are related to the fact that the part of the medial femoral condyle which articulates with the tibia is longer than that of the lateral femoral condyle. Active rotatory movements of the leg on the thigh are possible if the leg is bent to a right angle. These movements are impossible with the knee straight.

There are two fibrocartilaginous *menisci, medial* and *lateral*, in the knee joint (Fig.35e). They are attached to the top of the tibia and move with that bone. The lateral meniscus is nearly semicircular and the medial more elliptical. They are often called the *cartilages* of the knee and they are frequently torn, the medial more often than the lateral. When this occurs the torn part can move inwards and become wedged between the condyles of the femur and tibia with the result that the knee becomes locked in flexion. If the knee is manipulated so that the torn part is freed, movements at the knee become possible again. However locking of the joint can recur at any time and the best treatment is removal of the torn cartilage.

The patella has its own area of articulation with the femur and moves up and down in extension and flexion respectively. It is said to act like a pulley and improve the pull of the quadriceps femoris muscle, although if fractured it may be removed without much loss of function.

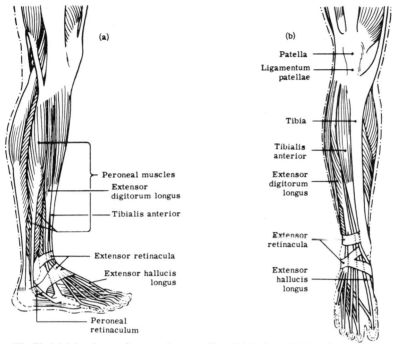

Fig.53. (a) Muscles on front and outer side of right leg, (b) Muscles on front of right leg.

The knee joint is very stable because of the large number of very strong ligaments between the bones. There are strong ligaments on the sides and back of the joint and also in the middle of the joint between the femur and tibia. The shape of the bone surfaces does not contribute very much to the stability of the joint. The upper surface of the tibia is flat and is made only slightly concave by the menisci. The lower end of the femur is convex especially at the back of the condyles.

THE MUSCLES OF THE LEG; THE ANKLE JOINT. The muscles of the leg are conveniently divided into an anterior group, a lateral group and a posterior group. The anterior group are attached to the fibula and the tibia and run on to the dorsum of the foot (Fig.53b). One goes to the inner border of the foot (the *tibialis anterior*), one to the great toe (the *extensor hallucis longus*) and one to the lateral four toes (the *extensor digitorum longus*).

The lateral group consists of two muscles (the *peroneus longus* and the *peroneus brevis*) which are attached to the fibula (Fig.53a). The

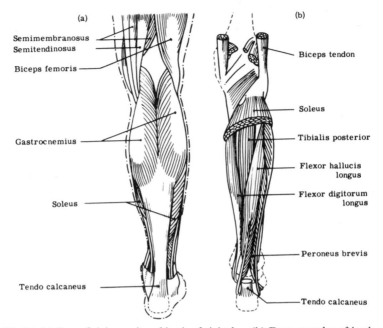

Fig.54. (a) Superficial muscles of back of right leg, (b) Deep muscles of back of right leg.

peroneus longus muscle runs downwards, winds round the lateral border of the foot and passes across the sole to be attached to its medial side. The peroneus brevis muscle is attached to the lateral border of the foot.

The muscles of the back of the leg are divided into two groups, a superficial and a deep. There are two large muscles, the *gastrocnemius* and *soleus*, in the superficial group (Fig.54a). The gastrocnemius muscle has two bellies in its upper part each attached to the back of the lower end of the femur. They can be easily seen when standing on tiptoe. The deeper muscle, the soleus, is attached to the tibia and fibula. These two muscles have a common tendon below, the *tendo calcaneus* known as *Achilles' tendon*, which is attached to the back of the heel bone, the calcaneus. The deep group of muscles is attached to the fibula and tibia and pass into the foot round the medial side of the ankle. One of them is attached to the medial side of the foot and two pass into the sole and go to the great toe (the *flexor hallucis longus*) and the lateral four toes (the *flexor digitorum longus*).

The ankle joint is a synovial joint of the hinge type. It is between the distal ends of the tibia and fibula above and the talus below

(Fig.32). At the joint the foot moves upwards and downwards, although the whole body moving on the foot is sometimes more important. Upward movement of the foot is called dorsiflexion (extension) and downward movement plantar flexion (flexion). The muscles producing dorsiflexion are the muscles on the front of the leg and passing into the foot. Plantar flexion is produced by the gastrocnemius and soleus muscles and may be assisted by the other deeper muscles. It is obvious that these large muscles do not simply move the foot downwards. They also lift the body upwards on the foot as in standing on tiptoe. These muscles are active in standing up from the sitting position, in walking and in going up stairs or up a slope. They pull the leg backwards on the foot fixed on the ground, or lift the body upwards on to the front of the foot. While standing they are continuously active since in the upright position the body tends to fall forwards at the ankle joints.

The ankle is a stable joint and is seldom dislocated. Its stability depends on the shape of the joint surfaces—the talus fits into the space between the malleoli and is held firmly there. In addition the medial, lateral and posterior ligaments hold the various bones together very firmly and the lower ends of the tibia and fibula are joined by a strong ligament just above the ankle joint. It is also true that fractures of the malleoli and tearing of the ligaments are common due to the enormous leverage exerted by the body on the ankle region if the body leans over too far to one side while the foot is fixed on the ground.

THE MUSCLES AND JOINTS OF THE FOOT. There is a small muscle on the dorsum of the foot. It goes to the medial four toes. In the sole there are several layers of muscles, some going from the heel to the toes, some entering the sole from the leg and some in the front part of the foot similar to the small muscles of the hand. The details of these muscles will not be given but some idea of their arrangement can be obtained from Fig.55a,b.

The hinge movements at the ankle joint have already been mentioned. The foot can also be turned inwards and outwards. The former movement is called *inversion* and in this movement the sole faces inwards and the inner border of the foot moves upwards and inwards. This movement is most easily performed when the foot is fully plantar flexed. The opposite movement, in which the sole of the foot is turned outwards and the outer border of the foot moves upwards and outwards, is called *eversion*. This movement is most

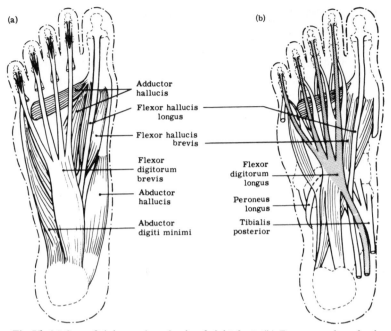

Adductor
hallucis

Flexor hallucis
longus

Flexor hallucis
brevis

Flexor
digitorum
brevis

Abductor
hallucis

Abductor
digiti minimi

Flexor
digitorum
longus

Peroneus
longus

Tibialis
posterior

Fig.55. (a) Superficial muscles of sole of right foot, (b) Deep muscles of sole of right foot.

easily performed when the foot is dorsiflexed. The main invertors are the tibialis anterior and tibialis posterior muscles and the main evertors are the peroneus longus and peroneus brevis muscles. Inversion and eversion are movements of the foot about the talus, that is the calcaneus (the heel bone) together with the rest of the foot swings inwards and outwards on the talus. These movements are extremely important in walking over uneven ground or when standing on one lower limb. They enable the body to move from side to side without straining the ligaments at the sides of the ankle joint or the malleoli forming the sides of this joint.

There is little or no movement between the rest of the tarsal bones and between the tarsal and metatarsal bones. There is, however, considerable dorsiflexion at the metatarsophalangeal joints. Again the importance of this movement is not the ability to turn one's toes upwards, which is due to the muscles going to the toes from the front of the leg, but the raising of the body on to the toes and finally the great toe as is seen at one stage while walking. The muscles responsible for this movement are the large calf muscles, the gastrocnemius and soleus.

There is some plantar flexion at the metatarsophalangeal joints. The muscles going to the toes from the back of the leg produce this movement. Side-to-side movements are also possible at the metatarsophalangeal joints but these are not important. Plantar flexion at the interphalangeal joints has a much greater range than dorsiflexion.

Before leaving the foot it is important to realize that the structure of the foot has been adapted for its function of weight-bearing and locomotion. Among the most striking differences as compared with the feet of the higher apes (and with the human hand) are the increase in length and size of the great toe and its loss of mobility. It lies parallel to the rest of the toes and its metatarsal

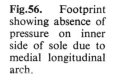

Fig.56. Footprint showing absence of pressure on inner side of sole due to medial longitudinal arch.

is very much larger in circumference as compared with the other metatarsals. The foot is also arched longitudinally from heel to toes especially on the medial side, and transversely in the region of the bases of the metatarsals. The typical shape of a footprint show this quite clearly (Fig.56). The arched, jointed foot gives a resilient structure which can yield to the force of the body weight in the same way as an arched bridge yields to traffic. An arch can be maintained in many ways (Fig.57) and the foot has a number of structures which help to maintain its arches. When these structures give way, as they sometimes do, fallen arches result. It should be added that

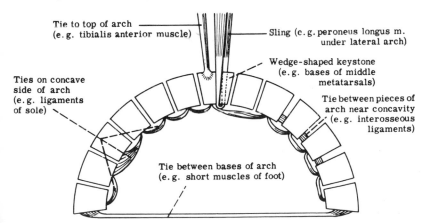

Fig.57. Ways in which an arch can be maintained.

while the arches are falling the foot is painful. A flat foot is not painful but functionally it cannot stand the stresses and strains which can be borne by the normal arched foot. The mobile arches of the ballet dancer's foot, which may look flat when weightbearing, are functionally more than satisfactory and show a normal arch when relieved of the weight of the body.

The fascia and synovial sheaths of the limbs

The deep fascia of the upper and lower limbs form special structures, some of which are very important. In the upper limb these are found in front of and behind the wrist and in the palm of the hand extending into the fingers (Fig. 58). On the front of the wrist there is a strong band of fascia passing between the medial and lateral carpal bones. It is called the *flexor retinaculum* and together with the underlying bones forms a tunnel through which the long flexor tendons and an important nerve pass into the hand. On the back of the wrist but somewhat higher up, a band of fascia, the *extensor retinaculum*, is attached to the lower ends of the radius and ulna (Fig.58a). It is attached to these bones in several places and forms a series of tunnels through which the tendons pass to the wrist and fingers. As the

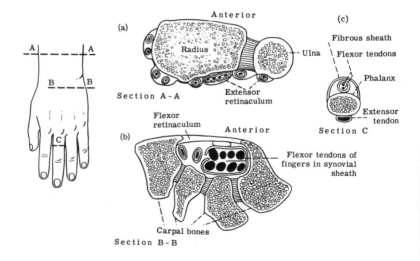

Fig.58. (a) Section through lower ends of radius and ulna (only extensor tendons are shown), (b) Section through distal row of carpal bones (only flexor tendons are shown), (c) Section through a finger.

flexor and extensor tendons pass into the hand under the retinacula they are surrounded by synovial sheaths which form a double layer round the tendon or tendons. Between the layers there is some synovial fluid which acts as a lubricant so that the tendon can slide up and down with a minimum of friction. The retinacula themselves hold the tendons down so that they do not bow when their muscles contract.

In the hand the *palmar aponeurosis* is deep to the skin to which it is firmly adherent. It is triangular with the base distal. Towards the fingers it divides into four slips which pass into the fingers and form tunnels by becoming attached to the sides of the phalanges. As an illustration of how strong fibrous tissue is, one may point out that if the palmar aponeurosis contracts due to disease the fingers become bent and remain bent. The muscles are quite incapable of straightening the fingers. The tendons in the fingers have synovial sheaths so that they can move easily in the canal formed by the bone and fascia (Fig.58c). These are found only in relation to the flexor tendons. They do not surround the extensor tendons. If the tendon becomes adherent to its synovial sheath due to infection or injury, bending the finger is impossible.

In the lower limb there is a strong stocking of deep fascia round the thigh and leg. This is attached to the hip bone and to the tibia at the knee as well as to the whole of the anteromedial surface of the tibia which can be felt in the leg. This fascia is especially thick on the outer side of the thigh and is called the *iliotibial tract*. The deep fascia has several thickenings round the ankle and the tendons which pass deep to these retinacula have synovial sheaths round them. The various retinacula are attached to bone on either side, form tunnels for the tendons and prevent bowing of the tendons when the muscles contract. The extensor and peroneal muscles each have two retinacula and the flexor one (Figs.53a,b).

5

The cardiovascular system

The heart

The heart and blood vessels are responsible for the circulation of the blood and may be regarded as a pump and a closed series of elastic, muscular tubes which branch and come together again so that the blood is returned to the heart. The heart is a muscular structure consisting of four chambers. It is cone-shaped with an *apex* below and to the left and a *base* above and to the right. It lies in the lower half of the thorax behind the sternum on the central tendon of the diaphragm.

THE POSITION OF THE HEART. It can be outlined approximately on the surface of the chest in the following way (Fig.59a). The right border is a line curved slightly to the right extending from the third right costal cartilage to the sixth right costal cartilage, both points 1 cm from the right edge of the sternum. The apex is in the fifth left intercostal space 9 cm from the midline. The left border is represented by a line from the apex to the second left costal cartilage 1 cm from the left border of the sternum. On either side of the heart is a lung enclosed in its pleura and behind the heart is the oesophagus.

THE PERICARDIUM. The heart is enclosed in a fibrous bag called the *fibrous pericardium* which is attached below to the diaphragm. Above, the edges of the mouth of the bag fuse with the walls of the large vessels entering and leaving the heart. Inside the fibrous pericardium, there is a double layer of *serous pericardium* (the outer is the *parietal* and is attached to the fibrous, and the inner is the *visceral* and is attached to the heart wall). The adjacent surfaces of the two layers are smooth so that they glide on one another.

88

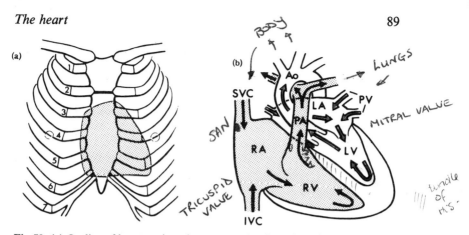

Fig.59. (a) Outline of heart projected on to anterior thoracic wall, (b) Interior of heart showing individual chambers and circulation of blood (SVC = superior vena cava, IVC — inferior vena cava, RA = right atrium, RV = right ventricle, PA = pulmonary artery, PV = pulmonary veins, LA — left atrium, LV = left ventricle, Ao = aorta).

THE STRUCTURE OF THE HEART. The heart is divided into right and left halves by a <u>muscular septum</u> which is oblique so that the left half is more posterior than the right (Fig. 60b). Each side of the heart is further subdivided into an upper and lower chamber called the *atrium* and *ventricle* respectively, so that there are *right* and *left*

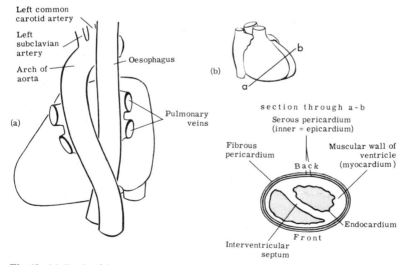

Fig.60. (a) Back of heart, showing aorta and oesophagus, (b) Section through ventricles showing comparative thickness of ventricular walls, oblique interventricular septum and structure of heart wall.

atria and *ventricles* (Fig.59b). The walls of the chamber consist of an outer covering of serous pericardium (*epicardium*), a middle layer of muscle (*myocardium*) and an inner layer of flattened endothelium (*endocardium*) (Fig.60b) The myocardium varies considerably in thickness. It is relatively thin in the atria and much thicker in the ventricles, and that of the left ventricle is twice as thick as that of the right. The atria are separated from the ventricles by a double ring of fibrous tissue marked on the outside by a groove called the *atrioventricular groove*. The fibrous rings surround the *atrioventricular openings*.

THE CHAMBERS OF THE HEART. The right atrium lies to the right and posteriorly, and forms the right border of the heart. Entering it vertically from behind are the *superior vena cava* above and the *inferior vena cava* below (Fig.59b). These large veins return the blood to the heart from the rest of the body. The blood supplying the heart itself returns to the right atrium by means of the *coronary sinus* which lies in the back part of the atrioventricular groove. The *interatrial septum* between the atria lies to the left, and the right atrioventricular opening, about 3 cm in diameter, is below and in front.

The right ventricle forms the right three-quarters of the anterior surface of the heart and is separated on the left from the left ventricle by the *interventricular septum* (Fig.60b).

Attached to the fibrous ring around the right atrioventricular opening and projecting into the ventricular cavity are the three flaps of the *right atrioventricular* (*tricuspid*) *valve*. The opposite edge of the flaps have cord-like structures attached to them (*chordae tendineae*) which in turn are attached to *papillary muscles* projecting from the wall of the ventricle (Fig.61a). The flaps of the valve prevent the blood passing back from the ventricle into the atrium and the chordae tendineae and papillary muscles prevent the flaps being turned inside out. Leading upwards from the right ventricle is the *pulmonary artery*. Just within the beginning of this artery is the *pulmonary valve* consisting of three cusps semilunar in shape and attached below with their free edge above. When the right ventricle contracts, it pushes the blood into the pulmonary artery. When the ventricle relaxes, the cusps come together and prevent the blood flowing back into the ventricle. (Fig.61b).

The blood in the pulmonary artery goes to the two lungs and the blood leaves the lungs in the *pulmonary veins*, two on each side.

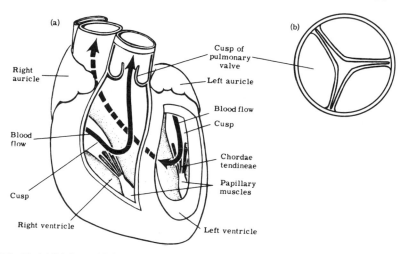

Fig.61. (a) Right and left ventricles opened from front to show structure of right and left atrioventricular valves, (b) View from above of three cusps of pulmonary valve.

These four veins pass horizontally to enter the *left atrium* which lies behind the upper part of the heart to the left (Fig.60a). To the right of the left atrium is the interatrial septum and below is the left atrioventricular opening, about 2 cm in diameter. Through this the blood passes into the left ventricle.

In the left ventricle, attached to the fibrous ring, are the two flaps of the *left atrioventricular* (*bicuspid, mitral*) *valve*. These flaps have chordae tendineae and papillary muscles. The flaps, etc., function in the same way as those of the right side of the heart. To the right is the interventricular septum. Leading upwards from the left ventricle is the *aorta* which lies behind and to the right of the pulmonary artery. Just within the beginning of the aorta are the three cusps of the *aortic valve*. This is similar to the pulmonary valve and prevents blood flowing back into the ventricle. Projecting forwards from each atrium on either side of the pulmonary artery is an *auricle* (*right* and *left*) (Fig.61a).

THE ARTERIES AND VEINS OF THE HEART. The heart is supplied with blood by the *right* and *left coronary arteries* which arise from the aorta just above its origin from the left ventricle (Fig.62a). The right coronary artery passes forwards between the right auricle and pulmonary artery and then winds round to the right in the

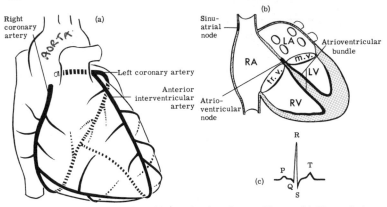

Fig.62. (a) Coronary arteries, (b) Conducting tissue of heart, (c) Normal electro-cardiogram (ECG).

P - Atrial systole
QRS - Ventricular systole
T - end of systole - repolarisation.

atrioventricular groove to the back of the heart. The left coronary appears between the left auricle and the pulmonary artery and winds round to the left in the atrioventricular groove to the back of the heart. The two arteries meet in this groove posteriorly.

Both arteries give off large branches. Both the main arteries and the large arteries have very few anastomoses with each other. The word *anastomosis* refers to connecting arterial channels between arteries before they break up into capillaries. These channels can be of great importance because if an artery is blocked the part supplied by it may obtain blood from an alternative source. Arteries which have no anastomoses are called *end arteries*, that is, they have no connexion with other arteries before they form capillaries. These are found in the grey matter of the cerebral cortex, the spleen, the kidneys and the lungs. The occlusion of a coronary artery or one of its large branches leads to sudden death. Partial occlusion leads to a condition called *angina pectoris* which is characterized by pain over the sternum on exertion.

The veins of the heart join the coronary sinus which opens into the right atrium. There are many small veins opening directly into the right atrium.

THE NERVE SUPPLY AND CONDUCTING TISSUE OF THE HEART. The nerve supply is from the autonomic nervous system and is both sympathetic and parasympathetic. The nerve fibres end in relation to the different parts of the *conducting tissue* of the heart (see below) and the heart muscle itself.

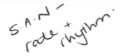

The *conducting tissue* of the heart refers to specially differentiated heart muscle. It consists of the *sinu-atrial (S-A) node*, the *atrioventricular (A-V) node* and the *atrioventricular (A-V) bundle (of His)* which ends in specialized cells called *Purkinje fibres*. The S-A node (called *the pacemaker of the heart*) lies to the right at the junction of the superior vena cava with the right atrium, and the A-V node lies in the atrioventricular groove posteriorly. The A-V bundle passes from the A-V node towards the interventricular septum where it divides into right and left branches which pass on either side of the septum towards the apex of the heart and then into the rest of the walls of the ventricles (Fig.62b). All these structures are like a syncytium, similar to cardiac muscle.

Normally the rate of the heart beat and its rhythmicity are controlled by the S-A node and the nerve supply of the heart. The S-A node sends out rhythmic impulses to the atrial muscle which conducts the impulses to the A-V node which in turn conducts the impulses to the ventricular muscle. In this way the contraction of the muscle of the whole heart is co-ordinated. The normal heart rate is about 70 beats per minute in a healthy adult at rest in reasonably warm surroundings. Stimulation of the sympathetic increases the heart rate and stimulation of the parasympathetic slows the heart rate. It should be noted that the heart without a nerve supply will contract spontaneously and rhythmically. If there is a failure of conduction from the atria to the ventricles along the conducting system the ventricles take on their own rhythmic rate which is much slower than that of the atria.

THE CARDIAC CYCLE. The *cardiac cycle* (0·8 sec) refers to the sequence of events which occur in the heart during one heart beat. This is divided into *diastole* (0·49 sec), a period during which the heart muscle is relaxed, and *systole* (0·31 sec) during which the muscle is contracted. The right and left sides of the heart behave in a similar manner during the cardiac cycle. In the first part of diastole blood flows into the atria through the veins (superior and inferior venae cavae to the right atrium and pulmonary veins to the left atrium). The filling of the atria results in the pressure within these chambers rising and the atrioventricular valves open so that blood flows into the ventricles. The atria and ventricles are filled and systole commences in the atria. This results in the stretching of the ventricles because the blood cannot flow back into the large veins due to their openings into the atria being closed by the muscle in their walls. The ventricles now contract, close the atrioventricular valves and force the blood into the aorta and pulmonary artery through the aortic and pulmonary valves. Before ventricular systole ceases blood is already flowing into the atria. With relaxation of the ventricles diastole commences again.

The cycle of events is accompanied by two *heart sounds* which can be heard through the chest wall. The first sound, described as *lŭbb*, is due to the closure of the flaps of the atrioventricular valves and therefore indicates the beginning of ventricular systole. The second sound, described as *dŭp*, of higher pitch and shorter and sharper than the first, is due to the closure of the aortic and pulmonary valves and indicates the beginning of ventricular diastole.

The *electrocardiogram* (ECG) is the record of electrical changes which occur in the heart muscle during the cardiac cycle. The ECG shows a series of waves which are labelled PQRST (Fig.62c). The P wave indicates, more or less, atrial systole, the QRS wave ventricular systole, and the T wave the end of systole. Changes in the ECG either in the wave form or in the length of time between its various parts may indicate disorders of the cardiac muscle and/or conducting system. The PR interval is particularly important in indicating delayed conduction in the A-V bundle, and should not be longer than 0·2 sec.

THE CARDIAC OUTPUT. The *cardiac output* is the amount of blood expelled from each ventricle per minute. This can be calculated by measuring the difference in the quantity of oxygen in the 100 ml of blood in the right atrium and 100 ml of arterial blood, and estimating the amount of oxygen removed by the lungs per minute (by means of a spirometer). By this means it can be shown that about 5000 ml of blood are expelled per minute. The *stroke volume* (the amount of blood expelled by the ventricle at each beat) is 5000/70 (number of beats per min), that is, about 70 ml. This is the approximate figure for a healthy adult at rest and can be increased considerably. Cardiac output can be increased to about 30 l per minute. This is achieved by increasing the heart rate, and the amount expelled with each beat is also increased. This is due to the increased venous return which results in the ventricle dilating to some extent and increasing its contraction. Increase in the heart rate has its limitations and if the heart beats too fast diastole is too short for proper filling and output falls.

Blood vessels

THEIR STRUCTURE. Vessels leaving the heart are called *arteries* and as these divide and become smaller they are called *arterioles*. The arterioles in turn become smaller until their wall is only one cell thick. These are called *capillaries* and they join up to form larger and thicker-walled vessels called *venules*. They in turn become larger and are then called *veins*. The blood is returned to the right atrium of the heart through the large veins called the superior and inferior venae cavae.

The structure of the different blood vessels varies considerably but all the arteries have a similar structure (Fig.63a). They consist of three coats, an inner called the *tunica intima*, consisting of flattened endothelial cells inside a layer of elastic fibres, a middle called the *tunica media* consisting of elastic tissue and smooth muscle, and an outer called the *tunica adventitia* consisting of fibrous tissue. In the largest arteries the tunica media consists almost entirely of elastic fibres and these are called *elastic arteries*. In the smaller arteries the tunica media consists largely of smooth muscle, circularly arranged. The transition of arteries to arterioles is mainly one of size—an arteriole has an outside diameter of less than 0·1 mm. Gradually the three coats of the arterioles become thinner and the outer two coats disappear so that a capillary consists of only endothelium (Fig.63c).

Venules consist of an endothelial lining and a few muscle fibres, and have a wider lumen and thinner wall than an arteriole. A vein

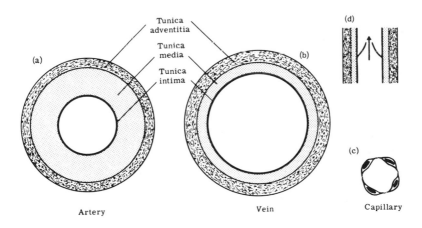

Fig.63. Transverse section of: (a) Artery, (b) Vein, (c) Capillary, (d) Longitudinal section of vein to show valve.

accompanying an artery has three coats but the wall is much thinner than that of the artery and the lumen is much larger (Figs.63a,b). Their tunica media is very much thinner and has less muscle than the artery, and the tunica adventitia is the thickest of the three coats. Most of the larger veins have *valves* consisting of two flaps. Each flap is a double layer of endothelium with a little connective tissue between the layers. The flaps project into the lumen and in the direction of the flow of blood so that they prevent the blood from flowing backward (Fig.63d).

The differences in the structure of the blood vessels are related to their functions. The elastic arteries, for example, the aorta, receive the blood from the left ventricle during systole and their walls are stretched. During ventricular diastole the pressure due to ventricular contraction falls and the walls of the elastic arteries recoil. The aortic valve is closed and the blood is pushed forwards due to this elastic recoil. The larger arteries apart from their elasticity also act as conducting channels.

THE ARTERIOLES. The blood has to be forced round the body in the blood vessels, and the heart has therefore to exert pressure on the blood. This is called *blood pressure* and is due to the force of the ventricular contraction. Blood pressure reaches its maximum during systole (*systolic pressure*) and its minimum during diastole (*diastolic pressure*). The difference between the two is called the *pulse pressure.* In a healthy adult at rest the systolic pressure is about 120 mm mercury and the diastolic is about 80 mm mercury. Blood pressure is an indication of the resistance in the peripheral vessels to the blood flow, and the greater the resistance, the greater the force required to pump the blood round the body, that is, the greater is the blood pressure. In old age the walls of the arteries lose their elasticity, that is, they harden, so that blood pressure rises with increasing age.

The muscular arteries and arterioles, because of the muscle in their walls, can contract and relax and thus vary the resistance due to the peripheral vessels. The blood pressure is at a maximum in the arteries nearest the heart and falls to about 40 mm mercury in the arterioles. This is due to the frictional resistance of the vessel wall and the increase in the surface area of the vascular bed. In the capillaries there is a further fall in the blood pressure (32 mm mercury at the arterial end and 12 mm mercury at the venous end) due to an increase in the surface area and cross-sectional area of the

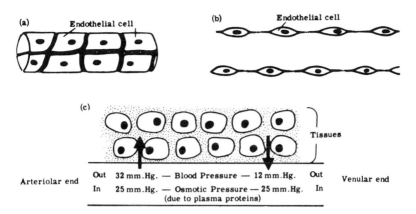

Fig.64. (a) Wall of capillary, (b) Longitudinal section of capillary, (c) Movement of fluid out of and into capillary.

vascular bed. The blood pressure falls still further in the venules and large veins (4 mm mercury). The venous return to the heart depends largely on pressure on the veins due to skeletal muscle contraction, and suction during respiration.

Blood pressure is increased if the arterial and arteriolar walls are constricted. The smooth muscle of the arteries and arterioles is always supplied by sympathetic nerves. Stimulation of the sympathetic nerves results in contraction of the muscle of the artery (the exceptions are the coronary arteries of the heart and the arterioles of striated muscle). Contraction of the vessel walls is called *vasoconstriction* and relaxation is called *vasodilatation*. The contraction and relaxation of the muscle of the walls of the arteries and arterioles not only influence the blood pressure but also control the amount of blood going to any structure of the body. These changes are regulated by *vasomotor centres* in the hindbrain.

Some examples of how this mechanism works may explain its importance. If the blood pressure falls, the impulses going to the vasomotor centres decrease. The vasomotor centres then send impulses to the sympathetic nerves of the blood vessels which constrict. This results in an increased resistance to the flow of blood, a more forceful contraction of the heart and a rise of blood pressure. A fall in the temperature of the blood acts on the vasomotor centre and results in a vasoconstriction of the arterioles of the skin so that less heat is lost. A rise in the temperature of the blood produces the opposite effect.

THE CAPILLARIES. The capillaries consist of a wall only one cell thick. They contain blood at a pressure of about 32 mm mercury at the arterial end and about 12 mm mercury at the venous end. The velocity of the blood in the capillaries falls because of the increase in cross-sectional area. The fall in blood pressure and in the velocity of the blood enables exchanges to take place between the blood and the tissue fluids. All the substances in solution in the blood, except the plasma proteins, can pass through the capillary wall (either between the cells or across the cell). At the arterial end the blood pressure is about 32 mm mercury which is counteracted by an osmotic pressure of about 25 mm mercury due to the plasma proteins so that there is a driving force out of the capillaries of about 10 mm mercury. At the venous end the osmotic pressure is 25 mm mercury and the blood pressure 12 mm mercury. The result is that at the venular end fluid and dissolved substances pass from the tissue fluids to the capillaries (Fig.64c).

THE VEINS. The veins are conducting channels and the blood returns to the heart due largely to the contraction of the body muscles squeezing the blood along the veins and to the suction effect of the thorax during respiration. The force of the heart also assists the return of the blood. Back-flow in the veins is prevented by the venous valves which are especially numerous in the veins of the lower limb. Venous return is decreased if the vascular bed is too dilated or the blood volume too small.

The main arteries

THE ARTERIES OF THE HEAD AND NECK. The *aorta* leaves the upper part of the left ventricle and arches backwards and to the left to reach the left side of the fourth thoracic vertebra. This part is called the *arch of the aorta.* It then runs down the left side of the thoracic vertebrae where it is called the *thoracic aorta* and gradually moves towards the midline. At the twelfth thoracic vertebra it passes behind the diaphragm and enters the abdominal cavity where it becomes the *abdominal aorta*. The arch of the aorta gives off, just above its origin, the two coronary arteries which supply the heart. From the convexity of the arch arise the *brachiocephalic* (*innominate*) *artery* (which is about 4 cm long and divides into the *right subclavian* and the *right common carotid arteries*), and the *left common carotid* and the *left subclavian arteries* (Fig.66a).

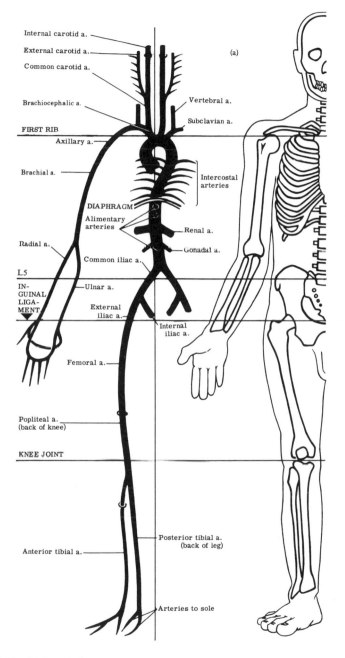

Fig.65a. Main arteries.

The subclavian artery gives off several important branches in the lower part of the neck and passes laterally over the first rib to become the main artery of the upper limb (*axillary artery*). Some of its branches in the neck enter the thorax, and some supply the thyroid gland and the neighbouring muscles. The *vertebral artery* is an important branch of the subclavian artery. It passes upwards in the foramen transversarium of the upper six cervical vertebrae and enters the skull through the foramen magnum. In the skull it supplies the posterior part of the brain (p. 200).

The common carotid artery passes upwards in the neck and at about the level of the fourth cervical vertebra divides into the *external* and *internal carotid arteries* (Fig.65a). The external carotid artery gives off a large number of branches which supply the structures in the neck and head outside the skull, for example, the thyroid gland, larynx, pharynx, tongue, face, scalp, external ear, muscles of mastication, walls of the nasal cavity, etc. The internal carotid artery enters the skull through a canal in the temporal bone and supplies the anterior two-thirds of the brain and the structures in the orbit. The thoracic aorta gives off the lower nine intercostal arteries and supplies branches to the lungs, diaphragm and oesophagus.

THE ARTERIES OF THE UPPER LIMB. The axillary artery lies in the armpit and continues into the upper arm as the *brachial artery* (Fig.65a). It lies on the medial side of the biceps brachii muscle and passes over the front of the elbow joint medial to the biceps tendon where it can be felt pulsating. An individual's blood pressure is measured over the brachial artery at this site. Just below this the brachial artery divides into the *radial artery*, which runs down the lateral side of the forearm, and the *ulnar artery* which runs down the medial side (Fig. 65a). The radial artery can be felt pulsating on the lateral side of the wrist on the radius. Just below this, the artery passes backwards behind the thumb and then forwards between the metacarpals of the thumb and index finger into the palm where it forms the main part of the *deep palmar arch* (Fig.65a). The ulnar artery enters the palm on its medial side and passes across to the lateral side forming the *superficial palmar arch*. Both arches give off branches which pass towards the digits.

THE ABDOMINAL ARTERIES. The abdominal aorta lies on the left of the bodies of the lumbar vertebrae and divides at the level of the fourth lumbar vertebra into the *right* and *left common iliac arteries*.

These pass downwards and laterally to the sacro-iliac joint where they divide into the *external* and *internal iliac arteries* (Fig.65a). The external iliac artery passes round the pelvic brim then deep to the inguinal ligament and becomes the artery of the lower limb (*femoral artery*). The internal iliac artery passes into the pelvis and supplies its walls and the pelvic viscera. The abdominal aorta gives off three large single arteries which supply the alimentary tract, liver, pancreas, and spleen. It also has branches going to the diaphragm, suprarenals, kidneys and gonads (testis or ovary).

THE ARTERIES OF THE LOWER LIMB. The femoral artery runs down the middle of the front of the thigh and then passes backwards medial to the femur and runs downwards behind the knee joint where it is called the *popliteal artery.* Beyond the knee joint it divides into the *posterior tibial* and *anterior tibial arteries* (Fig.65a). The posterior tibial artery continues down the back of the leg, then round the medial side of the ankle and into the sole of the foot, where it divides into two main branches which supply the digits. The anterior tibial artery passes forwards between the tibia and fibula into the front of the leg and downwards towards the foot. It passes between the two malleoli, enters the dorsum of the foot and ends by passing into the sole between the first and second metatarsal bones.

The main veins

THE VEINS OF THE LOWER LIMB. The veins of the limbs are both superficial and deep. In the lower limb there is a large superficial vein (*the great* or *long saphenous vein*) which begins on the medial side of the foot anterior to the medial malleolus (Fig.65b). It runs up the medial side of the leg, lies behind the medial condyles of the tibia and femur and continues upwards in the superficial fascia on the medial side of the thigh. It ends by going through a hole in the deep fascia (the *saphenous opening*) and joining the femoral vein. The *small* (*short*) *saphenous vein* runs upwards on the lateral side of the leg, passes to the back of the knee and perforates the deep fascia before joining the poplitial vein (Fig.65b).

The importance of the superficial veins of the lower limb cannot be sufficiently emphasized. The great saphenous vein is often used for transfusions where it lies in front of the medial malleolus at the ankle. Both veins frequently become varicose and are associated with discomfort, eczema and ulcers of the lower limb. The term *varicose*

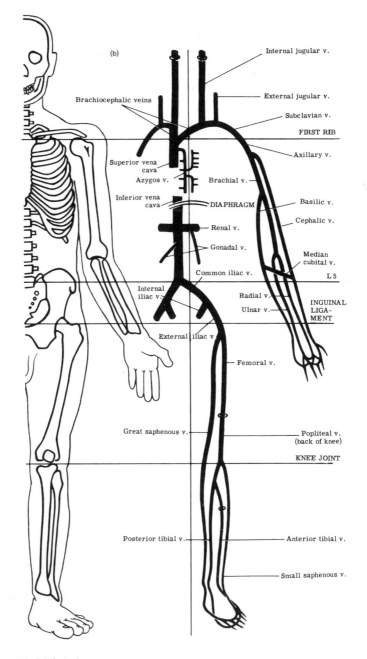

(b)

Internal jugular v.

Brachiocephalic veins

External jugular v.

Subclavian v.

FIRST RIB

Axillary v.

Superior vena cava

Azygos v.

Brachial v.

Inferior vena cava

DIAPHRAGM

Basilic v.

Renal v.

Cephalic v.

Gonadal v.

Median cubital v.

Common iliac v.

L 5

Internal iliac v.

Radial v.

Ulnar v.

INGUINAL LIGA-MENT

External iliac v.

Femoral v.

Great saphenous v.

Popliteal v. (back of knee)

KNEE JOINT

Posterior tibial v.

Anterior tibial v.

Small saphenous v.

Fig.65b. Main veins.

veins implies an enlargement, tortuosity and permanent distension of the veins. They are associated with incompetent valves.

The deep veins of the lower limb commence as veins in the sole and the deeper part of the leg. They form the *anterior* and *posterior tibial veins* which pass upwards towards the knee. The anterior goes posteriorly between the tibia and fibula, joins the posterior tibial veins and forms the popliteal vein which ascends to about halfway up the thigh, passes forwards, medial to the femur, and becomes the *femoral vein* (Fig.65b). The femoral vein enters the abdomen deep to the inguinal ligament on the medial side of the femoral artery. Both vessels are enclosed in fascia, *the femoral sheath*, which has three compartments (Fig.48). The artery is in the lateral compartment and the vein is in the middle. The medial compartment forms the *femoral canal*. Its importance is that a protrusion of peritoneum with some part of the abdominal contents, usually a loop of small bowel, may enter it and form a *femoral hernia*.

THE ABDOMINAL VEINS. The femoral vein becomes the *external iliac vein* and the veins of the pelvis join to form the *internal iliac vein* (Fig.65b). External and internal iliac veins join over the sacro-iliac joint and form the *common iliac vein*. The two common iliac veins unite to the right of the fifth lumbar vertebra and form the *inferior vena cava* which passes upwards a little to the right of the lumbar vertebrae and enters the thorax by passing through the right leaf of the central tendon of the diaphragm at the level of the eighth thoracic vertebra (Figs.44a,65b). This vessel has a very short thoracic course before entering the posterior part of the right atrium from below. The inferior vena cava in the abdomen receives veins from the gonads, kidneys, suprarenals, liver and diaphragm. The blood from the alimentary tract, spleen and pancreas passes to the liver in the *portal vein*. The portal vein breaks up into capillaries in the liver. By this means the blood from the alimentary tract is brought into close relationship with the cells of the liver. The blood from the liver goes to the inferior vena cava in the *hepatic veins* (Fig.66).

THE VEINS OF THE UPPER LIMB. The main veins of the upper limb are also superficial and deep. There is a medial superficial vein, the *basilic*, which begins on the medial side of the wrist and passes upwards on the medial side of the forearm then in front of the elbow where it receives a large tributary from the superficial vein of the lateral side of the forearm, the *cephalic vein* (Fig. 65b). The

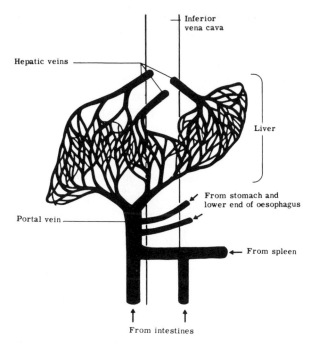

Fig.66. Hepatic venous portal system.

basilic vein ends by passing deeply in the upper arm about its middle and joins the deep vein in this region, the *brachial vein*. The cephalic vein begins on the lateral side of the wrist, passes upwards on the lateral side of the forearm and upper arm and ends by entering the axillary vein after passing through the deep fascia between the deltoid and pectoralis major muscles (Fig.65b). These superficial veins, especially the one joining the cephalic and basilic veins in front of the elbow (the *median cubital*) are used for obtaining specimens of blood and for intravenous injections.

The deep veins of the upper limb begin in the hand and form *radial* and *ulnar veins* which pass upwards in the front of the forearm. They join near the elbow and form the *brachial vein* which runs up the upper arm to the axilla (armpit) and becomes the *axillary vein* (Fig.65b).

THE VEINS OF THE HEAD AND NECK. The axillary vein forms the *subclavian vein* which joins the main vein of the head and neck, the *internal jugular vein*, to form the *brachiocephalic* (*innominate*) *vein* (Fig.65b). The *external jugular vein* receives a number of veins corresponding to many of the branches of the external carotid and subclavian arteries. This vein ends in the subclavian vein. The internal jugular vein receives the blood from the inside of the skull and some from the outside.

The *left brachiocephalic vein* crosses to the right above the arch of the aorta and passes slightly downwards. It joins the *right brachiocephalic vein* and forms the superior vena cava (Fig.65b). This vein passes vertically downwards for about 7 cm and enters the back of the right atrium from above. The veins of the thoracic wall end in the *azygos vein* which passes vertically upwards to the right of the vertebral column, arches forwards over the structures passing to and from the medial side of the right lung and ends in the superior vena cava (Fig.65b).

The veins inside the skull are called *venous sinuses*. The *superior sagittal sinus* runs backwards in the midline from the frontal bone to the occipital bone where it runs to the right to form the *right lateral sinus* (Fig.128). This leaves the skull through the jugular foramen to form the right internal jugular vein. The left internal jugular vein is formed in a similar manner but the *left lateral sinus* is formed by the *straight sinus* which runs backwards between the cerebral hemispheres and the cerebellar hemispheres (Fig.128). The straight sinus is formed by the union of the *inferior sagittal sinus* and the *great cerebral vein* which is formed by veins emerging from inside the brain itself.

THE FETAL CIRCULATION. The main differences between the fetal circulation and that of the newly born child are due to the fact that the blood in the fetus does not require to pass through the lungs and liver (Fig.67). The functions of these organs are performed by the mother. The oxygenated blood of the fetus returns from the placenta through the *umbilical vein* and instead of going to the liver goes to the inferior vena cava via a special vessel called the *ductus venosus*. The result is that the most highly oxygenated blood in the fetus enters the right atrium from below. It then passes through the septum between the two atria into the left atrium and then into the left ventricle from where the blood enters the aorta. Through the large main branches of the arch of the aorta, this highly oxygenated blood goes to the head, neck and upper limbs and returns to the right atrium through the superior vena cava. This blood, somewhat less oxygenated, passes into the right ventricle and then

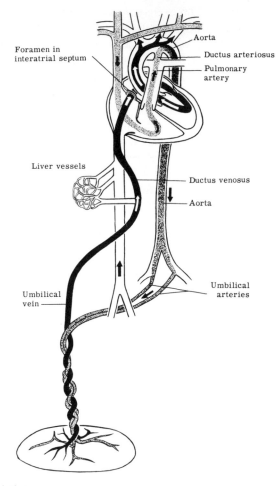

Fig.67. Fetal circulation.

into the pulmonary artery. It bypasses the lungs by going to the aorta through the *ductus arteriosus* which joins the pulmonary artery to the aorta beyond the origin of its large branches. The blood then goes to the rest of the body through the thoracic and abdominal aortae and their branches. It returns to the placenta through the umbilical arteries, continuations of the internal iliac arteries.

After birth the umbilical vein and ductus venosus are closed so that the inferior vena cava receives the blood from the lower limbs and the organs which, in the fetus, were only partially functioning (the kidneys and alimentary tract). This blood does not pass into the left atrium through the hole in the interatrial septum but goes into the right ventricle. The hole is closed because of the increase in pressure in the left atrium and the arrangement of a valve in the interatrial wall. All the blood now passes into the pulmonary artery and, due to the closure of the ductus arteriosus, goes to the lungs and returns to the left side of the heart through the pulmonary veins. The blood then circulates round the body through the aorta and its branches. The umbilical arteries are closed.

6

The blood and the lymph

The blood

Blood is the fluid tissue which circulates in the heart and blood vessels. It consists of a complex fluid in which solid elements (*blood corpuscles* or *cells*) are suspended. The cells form about 45 per cent of the volume of human blood, and the fluid about 55 per cent. There are about 6 litres of blood in the body.

If blood is prevented from clotting the corpuscles separate from the fluid part which is called *plasma* and consists largely of water (92 per cent) and proteins (7 per cent) in solution. The proteins are mainly *albumin* and *globulin*. Another important protein is *fibrinogen* which is necessary for clotting. One of the main functions of the proteins is the maintenance of osmotic pressure. Dissolved in the plasma are

   a. inorganic substances (0·9 per cent), such as sodium, potassium, calcium, magnesium, phosphorus, iodine and iron,
   b. additional organic substances such as urea, amino acids, neutral fats and glucose,
   c. hormones and antibodies.

If blood is allowed to clot, the clot separates and the fluid which remains is called *serum*.

THE BLOOD CORPUSCLES (CELLS). The corpuscles of the blood are

   a. red blood corpuscles (erythrocytes),
   b. white blood corpuscles (leucocytes),
   c. blood platelets (thrombocytes) (Table 2).

THE RED BLOOD CORPUSCLES. Red blood corpuscles are biconcave discs with a thickness at the edge of about 2 μm and a

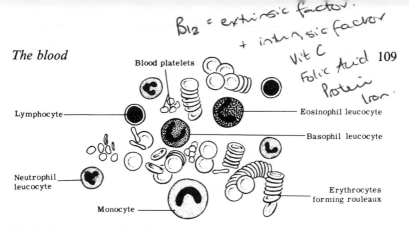

Blood platelets

Lymphocyte

Eosinophil leucocyte

Basophil leucocyte

Neutrophil leucocyte

Erythrocytes forming rouleaux

Monocyte

B₁₂ = extrinsic factor + intrinsic factor
Vit C
Folic Acid
Protein
Iron

Fig.68. Corpuscles (cells) of blood.

diameter of 7 μm (Fig.68). They have no nucleus. In a drop of fresh blood the red corpuscles form *rouleaux* in which the corpuscles resemble a heap of coins. Red corpuscles consist of a stroma or framework containing mainly water (about 60 per cent) and *haemoglobin* (30 per cent) which consists of an iron-containing pigment (*haem*) combined with a protein (*globin*). The main property of haemoglobin is its ability to form a loose compound with oxygen, although it also plays a part in regulating the acid–base balance of the blood and the carriage of carbon dioxide. The main function of the red corpuscles is the carriage of oxygen from the lungs to the tissues.

There are about 5 million red corpuscles per c.mm of blood in men and about 4½ million per c.mm in women. Red corpuscles are formed in the red bone marrow of the bones of the skeleton (mainly in the flat bones and bodies of the vertebrae) and after a life of about 110 days they are broken down in the spleen. The iron is stored in the liver and used again, and the pigments are used by the liver to form the bile pigments.

Several substances are required for the formation of a sufficient number of normal red corpuscles—protein, iron, vitamin C, vitamin B$_{12}$ (*extrinsic factor*), folic acid and an *intrinsic factor*, formed by the lining of the stomach. Liver contains both folic acid and the extrinsic and intrinsic factors. A common form of anaemia is due to lack of iron (*simple anaemia*). In this type the number of red corpuscles is fairly normal but the amount of haemoglobin in each corpuscle is reduced. The administration of iron will cure this form of anaemia. Another type of anaemia is called *pernicious anaemia* in which the number of red corpuscles is less than normal. This is much less common and is due to a lack of intrinsic and extrinsic factors, and the administration of liver will cure it.

THE WHITE BLOOD CORPUSCLES. White blood corpuscles contain a nucleus. They are divided into two types, *granular* and *nongranular*. The granular contain granules in their cytoplasm and because their nucleus is irregular in shape they are also called *polymorphonuclear* (*polymorphs*) (Fig.68). The number of lobes of the nucleus varies from two to five. The majority of the non-granular leucocytes are called *lymphocytes* in which the nucleus is spherical and large and there is very little cytoplasm. There are about 8,000 white cells per c.mm of blood and of these 25 per cent are lymphocytes. Most leucocytes are larger (10 μm in diameter) than red blood corpuscles.

Polymorphs are subdivided according to the staining properties of the granules. The majority (65 per cent of total leucocytes) are called *neutrophil*. These blood corpuscles can be actively amoeboid in their movements and are phagocytic to very small particles and organisms (streptococci and staphylococci), especially those which occur in acute infections. They also release substances which break down proteins (*proteolytic*). The *eosinophil polymorphs* (4 per cent) contain a large number of granules and are not amoeboid nor phagocytic. Their precise functions are not known but they are found in increased numbers in the blood as a reaction to the presence of foreign proteins in the body. *Basophil polymorphs* (1 per cent) contain differently staining granules and are amoeboid but not phagocytic. Their functions are not clearly known. The polymorphs are formed in the red bone marrow. They are thought to last about 12 hours in the blood and their fate is not known.

The lymphocytes are subdivided into *large* (10 μm) and *small* (7 μm). They are produced in lymphoid tissue throughout the body (lymph nodes, tonsils, spleen). Functionally they are associated with certain types of chronic infection, for example tuberculosis, and the production of antibodies. *Monocytes* (5 per cent) are also found in the blood. These are classified as non-granular leucocytes and are phagocytic in acute infections after the initial activity of the neutrophil polymorphs. They also ingest dust, carbon particles and organisms larger than those producing acute infections.

THE BLOOD PLATELETS. Blood platelets are about 2 to 3 μm in diameter and number about 250,000 per c.mm of blood. They do not contain any nuclear material. They are formed in the red bone marrow and functionally they are important in blood clotting.

Table 2

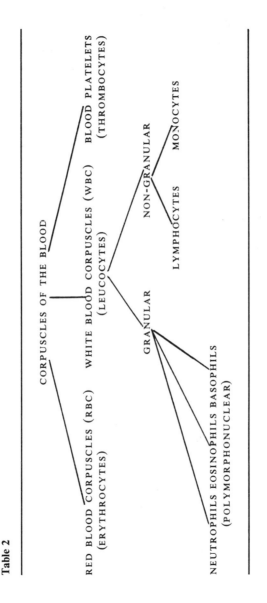

THE RED BONE MARROW. This tissue was referred to when bone was described. In the adult it is found in flat bones, for example the sternum, and irregular bones, for example the vertebrae, and in the proximal ends of the humerus and femur. The long bones contain yellow bone marrow, but in times of emergency the yellow marrow may be replaced by red. The red marrow consists of a sponge-like framework of fibres with cells between the fibres, and blood vessels. In this framework are the cells from which the red blood corpuscles, polymorphonuclear leucocytes and thrombocytes are developed. A specimen of red bone marrow, often obtained from the sternum by means of sternal puncture, shows every stage of these cells from the original cells from which they develop to the mature cells seen in the blood. The precursors of the red blood corpuscles are nucleated and in their very early stages of development do not contain haemoglobin. Polymorphonuclear leucocytes develop from the same cells (*haemocytoblasts*) as the red blood corpuscles. Blood platelets develop from a multinucleated cell (*megakaryocyte*).

Lymph, lymphatic vessels, lymphoid tissue

LYMPH. The fluid contained within lymphatic vessels is called *lymph*. The composition of lymph is very variable; for example, after a fatty meal the fat content is much higher. It also varies according to the site from which it is taken. Lymph resembles plasma but contains about half the proteins. It also contains lymphocytes in numbers similar to those found in blood.

THE LYMPHATIC VESSELS. The *lymphatic vessels* form a series of channels which pass from the tissues to the larger veins. They thus constitute an alternative route to the venous, for the return of fluid containing particulate matter, cells and dissolved substances, from the tissues. These vessels begin as blind-ending capillaries, join together to form larger vessels and ultimately form the main vessels which end in the brachiocephalic veins. Structurally lymphatic vessels are similar to veins. The capillaries consist of a single layer of endothelium and the larger vessels have three walls—intima, media and adventita—all of which are thin. The larger vessels have valves similar to those found in veins. Some tissues are not drained by lymphatic vessels, for example, bone, cartilage, and the central nervous system.

The lymphatic vessels of the lower limbs, abdomen, left side of the

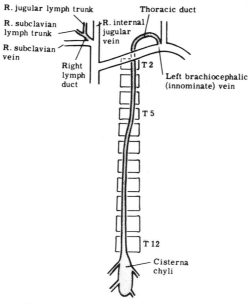

Fig.69. Thoracic duct.

thorax, left upper limb and left side of the head and neck drain into the *thoracic duct* which enters the beginning of the left brachiocephalic vein (Fig.69). The thoracic duct commences in the upper part of the abdomen below the aortic opening of the diaphragm. It passes upwards into the thorax through this opening and lies on the right of the thoracic vertebrae until it reaches the fourth or fifth thoracic vertebra where it crosses to the left behind the oesophagus. It continues upwards into the lowest part of the neck where it arches forwards to end in the beginning of the brachiocephalic vein. The *right lymphatic duct* is formed by the union of the *right jugular lymphatic trunk* (draining the right side of the head and neck) and the *right subclavian lymphatic trunk* (draining the right upper limb and the right side of the thorax) and ends in the beginning of the right brachiocephalic vein.

There are special lymphatic vessels in the villi of the lining of the small intestine. These are called *lacteals* because after a meal they appear as milky white streaks due to the fat in their interior.

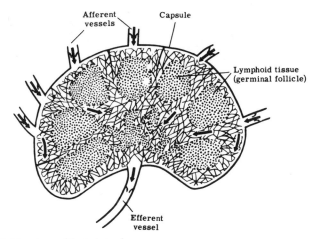

Fig.70. Structure of lymph node.

LYMPHOID TISSUE

Lymph nodes (*glands*). Lymphatic vessels, as they go towards the thoracic duct or right lymphatic duct, pass through several groups of lymph nodes. Lymph nodes, or glands as they are often called although they do not secrete anything, consist of aggregations of lymphocytes in a meshwork of fine fibres, surrounded by a fibrous tissue capsule which sends septa into the node (Fig.70). Several lymphatic vessels enter a node but only one leaves it. The lymph in the vessels percolates slowly through a number of thin-walled channels called *sinusoids* whose wall consists of a meshwork of fibres containing phagocytic cells which remove organisms, particulate matter, etc. from the lymph. Lymph nodes thus act as filters for the lymph which is within the vessels. Fluid from a tissue may pass through as many as three sets of nodes before returning to the veins. Lymphocytes are added to the lymph as it passes through the node. In addition to being a means whereby tissue fluid returns to the circulation and a source of lymphocytes, lymph vessels and nodes play a part in the spread of, and the defence against, infection and cancer.

It is important to know the sites of the main groups of lymph nodes. In the head and neck they are both superficial and deep. The superficial are arranged in a circle under the skin like a high collar (Fig.71a). There are groups

a. over the occipital bone just above the neck,

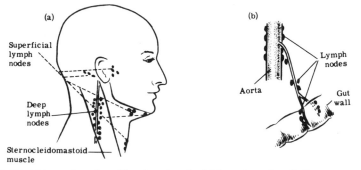

Fig.71. (a) Lymph nodes of head and neck, (b) Lymph nodes of alimentary tract.

b. over and in the parotid salivary gland in front of the ear,

c. below the body of the mandible,

d. below the chin,

e. in front of the larynx and trachea (windpipe).

The deep group lies along the internal jugular vein. The superficial drain into the deep and the deep into the jugular lymphatic trunk. Infection of the back of the scalp, for example, may cause enlargement of the nodes over the occipital bone or cancer of the tongue may spread to the nodes under the mandible and then to the nodes along the internal jugular vein.

The lymphatic vessels from the upper limb drain into lymph nodes in the armpit (axilla). There is usually a lymph node above the elbow joint. The axillary lymph nodes also receive vessels from the breast and front of the chest wall. The lymph nodes of the lower limb are found below the inguinal ligament and there are usually nodes at the back of the knee. The vessels from the lower part of the abdominal wall drain to the inguinal nodes.

The lymphatic vessels from the organs in the abdominal cavity pass through three groups of lymph nodes (Fig.71b).

a. near the organ,

b. on the vessels of the organ,

c. on either side of the abdominal aorta.

From the last group the vessels pass into the thoracic duct.

In the thorax there are three main groups of lymph nodes

a. a group near the large vessels in the' upper part of the thorax,

b. a group on the posterior thoracic wall,

c. a group round the trachea and its divisions, the main bronchi.

On the right side these groups join the right lymphatic duct and on the left the thoracic duct.

Other lymphoid tissue. Lymph nodes represent one form of lymphoid tissue which may be defined as aggregations of lymphocytes. These are usually held together by a capsule of connective tissue and a network of fine reticulin fibres. Besides lymph nodes, large masses of lymphoid tissue are found

a. in the spleen,

b. along the alimentary tract (at the back of the mouth, in the small intestine and in the appendix),

c. in the thymus.

The *spleen* is a very vascular organ lying in the upper left side of the abdominal cavity. It is about 15 cm long, 7·5 cm wide and 3 cm thick. Although it consists of masses of lymphoid tissue it has no lymphatic vessels, and the lymphocytes formed by this organ pass directly into the blood stream. The spleen will be further considered with the organs in the abdominal cavity. The lymphoid tissue of the alimentary tract is both defensive and a source of lymphocytes.

In recent years the *thymus*, an organ in the upper anterior part of the thorax, has been recognized as consisting of lymphoid tissue. It is large in children but atrophies after puberty. It is regarded as the source of the lymphoid tissue found in the lymphoid organs of the rest of the body and therefore the source of the lymphocytes which are concerned with immunity. Subsequently lymphocytes produced by the bone marrow are processed by the thymus and form the T-lymphocytes which are responsible for producing the reaction rejecting foreign tissue, e.g. a graft. The lymphoid tissue of the gut produces B-lymphocytes (so called because they are produced by the Bursa of Fabricius in the abdomen of the chicken). The B-lymphocytes develop into plasma cells which are responsible for the production of circulating antibodies found mainly in the gamma globulin of the plasma proteins.

7

The respiratory system

The respiratory tract includes the nasal cavity, pharynx, larynx, trachea, main bronchi and lungs. The mouth may be used in respiration but is really part of the alimentary tract. The upper and lower respiratory tracts lie above and below the division of the trachea respectively.

The nasal cavity

The first part of the cavity is surrounded by the *external nose* whose framework consists of bone above (mainly the nasal bones) and cartilage below.

The nasal cavity is divided into two by the vertical *nasal septum* which lies approximately in the midline in the sagittal plane. The septum is bony behind and cartilaginous in front. Each half has a roof, a floor (much wider than the roof), an outward sloping lateral wall and a medial wall (one side of the nasal septum). The roof is horizontal in its middle part and slopes downwards in front and behind. The horizontal part is formed by a plate with small holes in it and through these holes pass the central processes of the first neurones of the sense of smell. The floor is formed by the hard palate in front (two-thirds) and the soft palate behind (one-third) (Fig.72a).

The lateral wall is irregular because of three projections which pass medially—horizontally at first and then downwards (*inferior middle* and *superior concha*). Between each concha and the wall is a space called the *meatus* (*inferior, middle* and *superior meatus* below the respective concha (Fig.72b)). Opening into the inferior meatus is the *nasolacrimal duct* which passes downwards from the *lacrimal (tear) sac* on the front of the medial side of the orbit. Excess tears

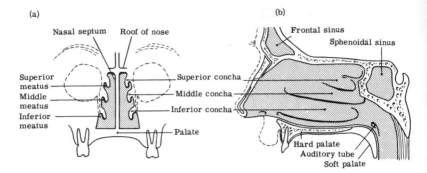

Fig.72. (a) Coronal section through nasal cavity, (b) Lateral wall of nasal cavity.

to some extent pass into the front of the nasal cavity. Everybody is familiar with the sniffing which accompanies excess production of tears.

THE PARANASAL AIR SINUSES. There are air spaces, called *air sinuses*, in the bones around the nasal cavity. Above is the *frontal sinus*, one on each side of the midline in the frontal bone; between the nasal cavity and the orbit are many small spaces; and in the maxilla is a large space called the *maxillary sinus* (Fig.73). This is the one usually referred to in the condition called *sinusitis*. All these sinuses communicate with the nasal cavity, most of them with the middle meatus. Their lining is continuous with the lining of the nasal cavity and has the same structure. There is also an air space in the

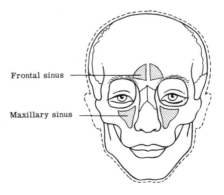

Fig.73. Position of frontal and maxillary air sinuses.

sphenoid bone behind the upper part of the nasal cavity and it opens into the back of the upper part of the cavity. All the sinuses are small at birth and usually enlarge to some extent between six and seven years. They all show considerable enlargement at puberty.

Functionally all the sinuses are said to add resonance to the voice and make the head lighter. There is little evidence that air enters and leaves them during respiration so that it is unlikely that they warm the inspired air. In lower animals there is considerable evidence that the sinuses may have an olfactory and/or a temperature-regulating function. Apparently they persist in man but have largely lost their original functions.

THE OPENINGS OF THE NASAL CAVITY. The nasal cavity opens on to the face by the *nares (anterior nasal apertures)*. They usually face downwards and are narrower in front than behind. They are surrounded entirely by cartilage. Just inside the anterior opening is a dilatation which is lined with skin on which are coarse hairs called *vibrissae*. These curve downwards and forwards and may arrest particles in the inspired air. The nasal cavity opens posteriorly into the nasopharynx by the vertically placed *posterior nasal apertures*. Each is oval measuring about 2·5 cm vertically and 1·5 cm transversely.

THE STRUCTURE AND FUNCTIONS OF THE NASAL CAVITY. The nasal cavity is lined by a mucous membrane covered by ciliated columnar epithelium. The mucous membrane is adherent to the underlying periosteum. It is continuous with the mucous membrane of the nasopharynx and the air sinuses which open on to the cavity. The mucous membrane is thick and vascular over the conchae and nasal septum, but thin in the meatuses and floor. There are many goblet cells and serous and mucous glands in the lining of the cavity. These keep the surface moist and also moisten the inspired air. The vascularity of the lining warms the inspired air and the mucus traps dust in the air. This is passed backwards towards the nasopharynx by the cilia.

There are two types of sensory nerves supplying the nasal cavity, general and special. The general nerves come almost entirely from the maxillary nerve, and the special are the olfactory nerves. The smell area of the cavity is on the roof and the adjacent areas of the septum and lateral walls.

The functions of the nasal cavity are associated mainly with

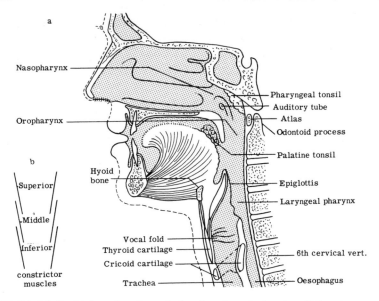

Fig.74. (a) Sagittal section through head and neck showing different parts of pharynx, (b) Arrangement of constrictor muscles of pharynx.

respiration, to take in and expel air. It also warms and filters the air. The sense of smell is found in the nasal cavity which is also used in speech for the pronunciation of nasal consonants (n, m, ng).

The pharynx

The *pharynx* lies in front of the vertebral column and behind the nasal cavity (*nasopharynx*), the mouth (*oropharynx*) and the entrance to the larynx (*laryngeal pharynx*). It consists largely of muscle, is tubular and extends from the base of the skull to the level of the sixth cervical vertebra where it becomes the oesophagus. (The cricoid cartilage marks this change on the front of the neck.) It is attached to the base of the skull, the mandible, the palate and the skeleton of the larynx (Fig.74).

THE NASOPHARYNX. The nasopharynx becomes the oropharynx at the level of the soft palate. It is roofed over by the base of the skull (mainly the occipital bone) and the atlas is behind it. In front are the posterior nasal apertures through which it communicates with the nasal cavity. In the lateral wall on each side just behind the nasal cavity at the level of the inferior meatus is the medial end of the *auditory (Eustachian) tube*. There is a mass of lymphoid tissue in the posterior wall of the nasopharynx (*pharyngeal tonsil*) which when enlarged is called the *adenoid*.

THE OROPHARYNX. The oropharynx is continuous above with the nasopharynx at the level of the soft palate, and below with the laryngeal pharynx behind the tongue. The axis and third cervical vertebra lies behind it (Fig.74a). In front it is continuous with the mouth through the *isthmus of the fauces* which is bounded above by the soft palate, below by the tongue and on each side by the *palatoglossal arch*, a fold passing between the soft palate and the tongue. Behind this arch and nearer the midline is the *palatopharyngeal arch* which passes from the soft palate to the pharynx. Between the two arches is the *palatine tonsil* (usually referred to as the *tonsil*) consisting of lymphoid tissue (Fig.74a). On its medial (free) surface are many *crypts* which are narrow recesses leading into the substance of the tonsil. The tonsil extends into the neighbouring tissues to a greater or less extent, for example into the soft palate above and tongue below. If enlarged, the tonsils may meet across the midline. The tonsils together with the pharyngeal tonsil of the nasopharynx and the lymphoid tissue on the back of the tongue (*lingual tonsil*) complete a ring which protects the openings into the respiratory and alimentary tracts.

THE LARYNGEAL PHARYNX. The laryngeal pharynx communicates above with the oropharynx and below with the oesophagus. Behind it are the fourth, fifth and sixth cervical vertebrae. In front is the opening into the larynx.

THE STRUCTURE AND FUNCTIONS OF THE PHARYNX. The pharynx is essentially a muscular tube extending from the base of the skull to where it is continuous below with the oesophagus. Its main three muscles are called the *constrictors* and they are related to each other in the same way as three plant pots are when placed inside each other (Fig.74b). Although air passes up and down the pharynx during respiration, the pharynx is also an important part of the alimentary tract and its arrangement of muscles is such that food is gripped by the upper constrictor muscles and pushed down into the lower part of the pharynx and then into the oesophagus.

The pharyngeal muscular tube, however, is deficient in front where it lies behind the nasal cavity, mouth and larynx, and the pharynx has important functions in preventing food from passing upwards into the nasopharynx and thence into the nasal cavity, from re-entering the mouth and from entering the laryngeal opening. It is assisted in these functions by other structures. The soft palate can be elevated and moved backwards so that the nasopharynx is shut off from the oropharynx. The arches on either side of the fauces come together to shut off the oropharynx from the mouth, and the whole larynx is lifted up behind the tongue so that its opening is closed and food passes behind the larynx into the oesophagus. All these muscular movements occur during swallowing. Besides the constrictor muscles there are other muscles of the pharynx which help to perform these movements.

One last function of the pharynx should be mentioned. The auditory (Eustachian) tube passes from the nasopharynx to the middle ear and by this means air gets into the cavity of the middle ear. The tympanic membrane (ear-drum) separates the middle ear from the external ear, and it is important that the air pressure on each side of the ear-drum is equal. This is ensured by the Eustachian tube which is opened during swallowing. If the tube becomes blocked the air in the middle ear is absorbed and changes occur in the middle ear and ear-drum so that one form of deafness occurs. Most people are familiar with the effect on the ears while flying in an aeroplane. The pressure outside the ear-drum is reduced and swallowing equalizes the pressure on the inner side of the drum, that is, in the middle ear. Infection can also spread from the pharynx to the middle ear along the Eustachian tube.

The larynx

The larynx is the organ of the voice as well as part of the respiratory tract. It extends from the back of the tongue to the trachea with which it is continuous at the level of the sixth cervical vertebra. It lies in front of the laryngeal pharynx and its lining is continuous with that of the pharynx above and the trachea below. In an adult it is about 4·5 cm long, 4·0 cm transversely and 3·5 cm from front to back. It is smaller in the female than in the male, particularly in its front-to-back diameter. This change in size occurs at puberty and is associated with the breaking of the voice. Before puberty the larynx in the female is about the same size as that in the male.

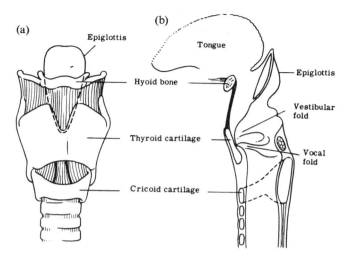

Fig.75. (a) Skeleton of larynx, (b) Sagittal section through larynx.

THE STRUCTURE OF THE LARYNX. The larynx consists of a skeleton of cartilages held together by ligaments and membranes (Fig.75a). It also has a number of small muscles which move the cartilages relative to each other. Some of these muscles arc important in closing the larynx during swallowing. The hyoid bone should also be included in the skeleton of the larynx. It lies in the upper part of the neck and can be felt quite easily below and within the angles of the mandible. Below the hyoid bone is the *thyroid cartilage* which consists mainly of two plates which meet in front in the midline. Where they meet there is a projection much more marked in men than in women. This projection is known as *Adam's apple* or the *laryngeal prominence*. Below the thyroid cartilage the ring-like *cricoid cartilage* can be felt. Below this the larynx is continuous with the trachea. The *epiglottis* is a leaf-like piece of cartilage which lies behind the tongue in the midline and projects upwards behind the hyoid bone.

The larynx is lined by a mucous membrane which is continuous with the lining of the pharynx above and that of the trachea below. It consists of ciliated mucous columnar epithelium and is typical of the lining of the major part of the respiratory tract, including the nasal cavity. Inside the larynx, at the level of the laryngeal prominence, are two horizontal folds running more or less from the front to the back of the larynx. They are called the *vocal folds* or *true vocal cords* and they are the important structures in producing what are known as voiced sounds as opposed to whispering. The *false vocal cords* (*vestibular folds*) are above the true vocal cords and are not used in speech (Fig.75b).

THE FUNCTIONS OF THE LARYNX. It maintains an open passage for air in respiration and to some extent influences the amount of air entering and leaving the lungs by altering the cross-sectional area of the larynx. This is brought about by approximating or separating the vocal folds by the action of the muscles which move the cartilages at the back of the larynx to which the vocal folds are attached.

It can also close the air channel when necessary. This occurs during swallowing when the entrance to the larynx is closed and the vocal folds meet in the midline. Closure of the opening is also a necessary precursor to forced expiration as in coughing, so that the pressure below the closure is built up. Strong muscular effort also requires a closed larynx so that the muscles used in the effort can act without their moving the chest wall.

The larynx provides a cleansing surface for the inhaled air. Particles are caught in the mucus and this is moved upwards by the cilia towards the pharynx.

The pitch and loudness of the voice depend on the vocal folds. By means of muscles the functioning length of the folds, their tension and thickness can be changed and thus determine the quality of the voice. High-pitched sounds, for example, are produced by making the folds short, thin and tense.

The trachea (Fig.76)

This is about 10 cm long and about 2 cm wide and extends from the sixth cervical vertebra to the fifth thoracic vertebra where it divides into the *right* and *left main bronchi*. It lies mainly in the midline but where it divides it lies slightly to the right. In outline it is more or less circular but is flattened behind. In the neck it is relatively superficial and its rings can be felt above the manubrium sterni. In the thorax it lies much more deeply behind the arch of the aorta. Behind it, along its whole length, is the oesophagus. On either side are the large vessels of the neck. The isthmus of the thyroid gland crosses the second, third and fourth tracheal rings.

There are about twenty incomplete rings of cartilage in the trachea. The rings are defective posteriorly. The rest of the wall

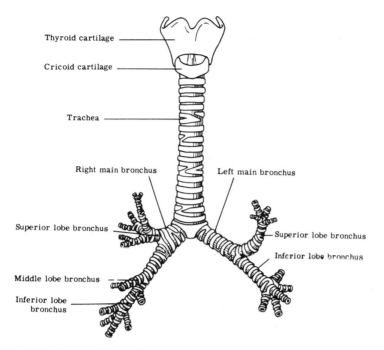

Fig.76. Trachea, main bronchi, lobar bronchi and segmental bronchi.

consists of connective tissue and smooth muscle and the trachea is lined by a ciliated, mucous, columnar epithelium. The walls of the trachea are elastic to some extent and both its length and width can be varied; for example, on inspiration the trachea is lengthened. The main function of the trachea is to provide a passage for air during respiration and the rings maintain its patency.

The main bronchi

The right main bronchus passes to the hilum of the right lung and is more vertical than the left (Fig.76). The right bronchus, about 2·5 cm long, is accompanied by the right pulmonary artery and the right pulmonary veins. Before entering the lung the right main bronchus gives off a branch to the upper lobe of the right lung and then divides into two. The upper branch goes to the middle lobe and the lower to the lower lobe of the lung. The left main bronchus is about twice as long as the right and passes below the aortic arch. The left pulmonary artery and the left pulmonary veins accompany it. The main bronchus divides into two, one branch going to the upper lobe and one to the lower lobe of the left lung.

The structure and functions of the bronchi are similar to those of the trachea.

The mediastinum

This is the name given to the space in the middle of the thorax between the two lungs and pleurae. It extends from the sternum in front to the vertebral column behind, and from the thoracic inlet above to the diaphragm below. It is divided into a *superior* and *inferior mediastinum* (above and below the manubriosternal joint). The superior mediastinum contains the large vessels, the vagus and phrenic nerves, and the trachea and oesophagus. The inferior mediastinum contains the heart, the beginning or end of the large vessels and the oesophagus behind the heart.

The pleurae and lungs

THE PLEURAE. These are double folds of serous membrane covering the lungs. Each lung is covered by its own pleura whose two layers are continuous with each other at the hilum of the lung. The outer *parietal pleura* is adherent to and lines the thoracic wall and covers the upper surface of the diaphragm. The *visceral pleura* is adherent to and covers the lung, and extends into its fissures. There is a potential space between the two layers (*pleural cavity*), and normally the parietal and visceral pleurae are in contact (Fig.77).

The pleura consists of loose connective tissue, but the cells of the adjacent surfaces of the two layers are flattened. There is a capillary layer of fluid between the parietal and visceral pleura. This enables the layers to glide on each other although they are separated with difficulty (compare two moistened glass slides placed together).

The subatmospheric pressure in the pleural cavities (754 mm mercury in inspiration and 758 in expiration) is a very important factor in enabling respiration to take place. A pressure in the pleural cavity equal to or greater than that of the atmosphere makes respiration difficult or impossible. The pleural pressure is also an important factor in maintaining the venous return to the heart.

THE LUNGS

The external appearance of the lungs. The lungs are the essential

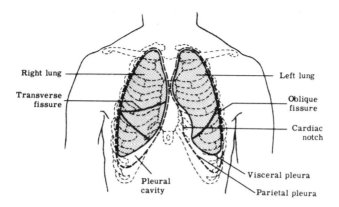

Fig.77. Surface markings of pleurae and lungs.

organs of respiration. They are cone-shaped with an *apex* superiorly, a concave base, a convex costal surface (posterior, lateral, and anterior) and a flattened medial surface (Figs.77, 78). Each lung occupies nearly one-half of the thoracic cavity and is covered by its pleura. In adults the surface is mottled due to the deposition of carbon particles in the tissues below the surface. This mottling is not seen in the newborn infant and is less marked in people who have lived in the country.

The costal surface is related to the ribs behind, laterally and in front. The apex, because of the sloping nature of the thoracic inlet extends about 3 to 4 cm into the neck above the medial end of the clavicle. The base (diaphragmatic surface) is related to the diaphragm. Below the right lung and right half of the diaphragm is the

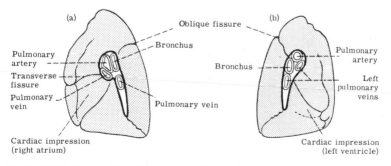

Fig.78. (a) Mediastinal (medial) surface of right lung, (b) Mediastinal surface of left lung.

liver, and below the left lung and left half of the diaphragm are the stomach and the spleen. About the middle of the medial surface there is a depression called the *hilum* where the pulmonary artery, pulmonary veins and bronchus (forming the *root* of the lung), enter and leave the lung. The heart is related to this surface below the hilum. The posterior part of the medial surface is massive and rounded, and lies in the hollow lateral to the thoracic vertebral column.

There are two important differences between the right and left lungs. The anterior border of the left lung has a notch (*cardiac notch*) below the level of the fourth costal cartilage. This notch deviates to the left for about 4 cm. Secondly, the right lung has three lobes and the left has two. Both have an *oblique fissure* which divides the lungs into *superior* (*upper*) and *inferior* (*lower*) *lobes*. The lower part of the superior lobe is anterior to the upper part of the inferior lobe (Fig.78). In addition in the right lung there is a *horizontal fissure* dividing the superior lobe into superior and *middle lobes*. The fissure is at the level of the fourth costal cartilage and is seen anteriorly and medially (Figs.77, 78). Each lobe is subdivided into *bronchopulmonary segments* each functionally independent in that it has its own bronchus and pulmonary artery and vein.

The *root* of the lung refers to the structures entering and leaving the hilum. In each root there is a bronchus, a pulmonary artery and two pulmonary veins.

The structure of the lungs. The microscopic structure of the lung is best described in terms of the air passages dividing and subdividing into smaller and smaller branches accompanied by smaller and smaller branches of the pulmonary artery. As the bronchi get smaller

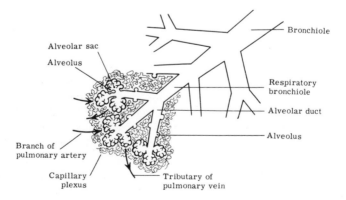

Fig.79. Terminal divisions of respiratory passages.

they are called *bronchioles*. Finally the bronchioles divide into several *respiratory bronchioles*. A respiratory bronchiole divides into *alveolar ducts* which further subdivide into *alveolar saccules* at the ends of which are the *alveoli* (Fig.79). The pulmonary arteries end as a capillary network around the alveoli. From this capillary plexus vessels arise which form the pulmonary veins. These veins run towards the hilum independently of the bronchi and pulmonary arteries.

Changes take place in the structure of the air passages as these divide and subdivide. The bronchi are lined by ciliated, mucous, columnar epithelium and have a wall consisting of fibrous tissue containing many elastic fibres, cartilage and smooth muscle. The air is cleansed by means of the mucus and cilia but when it has reached the alveoli the presence of mucus and a thick wall would interfere with the passage of oxygen and carbon dioxide through the alveolar wall. One can appreciate why the epithelium of the alveolus is flat (squamous) and has no mucous cells. The rest of the wall has disappeared and there is no cartilage or muscle round the alveolar saccules. Higher up the respiratory tract the cartilage keeps the lumen patent and the muscle regulates the size of the lumen. Elastic fibres are found everywhere because the elasticity of the lung tissue is extremely important in the expansion and contraction of the lung during respiration.

The functions of the lungs. The main functions of the lung are to supply the body with oxygen and get rid of excess carbon dioxide. In so far as expired air is usually warmer than inspired air and is saturated with water vapour normal respiration is associated with some heat loss and some water loss. In quiet respiration about 500 ml of air are breathed in and out in each respiratory cycle. If the respiratory rate is about 18 per minute the volume of air entering and leaving the lungs is about 9 l per minute. This can be increased by deeper and more rapid breathing to 150 l per minute. Slow, deep breathing is much more efficient than rapid, shallow breathing for increasing the amount of oxygen available to the body.

Expansion and contraction of the lungs (inspiration and expiration) are due to the activity of the respiratory muscles, the diaphragm, etc. This activity is regulated by respiratory centres in the hindbrain. These centres are affected directly by the amount of carbon dioxide in the blood. A rise, for example as a result of exercise, leads to deeper and later more rapid breathing and a fall to more shallow and slower breathing.

Lack of oxygen affects breathing in two ways. Directly, the result is a failure of function of the respiratory centres, but indirectly, sensory impulses from the carotid body (near the bifurcation of the common carotid artery) stimulate the inspiratory centre and breathing becomes deeper and more rapid. There are similar receptors on the aorta.

An increase in the temperature of the blood acts directly on the respiratory centre and quickens breathing. Emotional influences, such as fear and excitement, and stimuli from the skin affect breathing. Breathing is also affected by proprioceptive impulses from muscles and joints. In some activities, for example, in speech and singing, breathing is controlled. Breathing is also altered in yawning and sighing, irritation of the respiratory passages may result in sneezing or coughing, and in swallowing respiration is inhibited.

8

The alimentary system

Basically this system, which includes the alimentary tract and the glandular structures associated with it (the salivary glands, liver and pancreas), is responsible for the intake, breakdown and assimilation of food. The upper part, however (the lips, tongue, mouth, etc.), is also involved in respiration and speech.

The oral cavity

THE LIPS AND CHEEKS. The lips are two fleshy folds surrounding the opening of the mouth. Each lip consists mainly of muscle covered by skin externally and mucous membrane internally. Both are bound down to the underlying tissue. The red margin of the lips is seen where the skin meets the mucous membrane. Both are covered by stratified squamous epithelium but the epidermis of the skin is keratinized and hairy, whereas the epithelium of the mucous membrane is non-keratinized and contains mucous and salivary glands. The keratin largely explains the difference in colour of the two parts of the lips. The upper lip has a vertical groove in the midline externally (*philtrum*). The *angle* of the mouth is where the two lips meet laterally. The *frenulum* is a vertical fold of mucous membrane in the midline between the inner surface of each lip and the gum. The upper is much more marked than the lower.

The lips are usually closed in chewing and swallowing, are used in sucking and blowing and have an erotic function. They play a very important part in speech. In association with the last-named function the lips have very large cortical areas, both sensory and motor, relative to their size in comparison with other parts of the body.

The cheeks have a structure similar to that of the lips—an outer skin, an inner mucous membrane and a layer of muscle between.

The parotid duct opens on the inside of the cheek opposite the upper second molar tooth. Both the cheeks and lips have a very rich blood supply from branches of the external carotid artery, for example, the facial artery. The cheeks are used in chewing—they push the food between the teeth. They are used also in blowing, sucking and speech.

THE MOUTH. This cavity consists of a smaller outer part between the cheeks and teeth and a larger inner part. With the teeth closed the outer part communicates with the inner by a space behind the molar teeth. The parotid duct opens into the outer part opposite the upper second molar tooth.

The inner part is surrounded laterally and in front by the gums and teeth. Behind it leads into the oropharynx through the *isthmus of the fauces* which is bounded by vertical folds. The mouth is

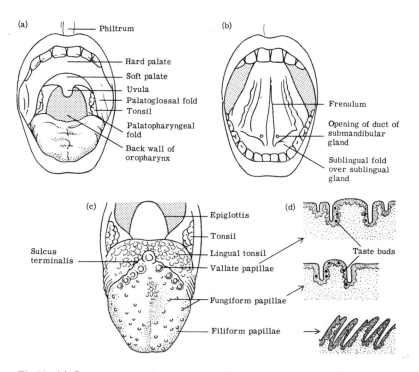

Fig.80. (a) Structures seen inside mouth, (b) Floor of mouth after elevating tongue, (c) Upper (dorsal) surface of tongue pulled well forwards, (d) Papillae of tongue.

roofed over by the hard and soft palates which separate it from the nasal cavity. The floor of the mouth is formed by muscles meeting in the midline. In the floor is the tongue. The inferior surface of the tongue is joined to the floor by a vertical midline fold (*frenulum*). On either side of the frenulum is the opening of the submandibular duct and a ridge extending laterally is seen between the tongue and the floor of the mouth. This ridge is produced by the underlying sublingual salivary gland (Fig.80b).

The mouth is used in chewing and swallowing. Alteration of its size and shape during speaking results in different sounds being produced.

THE SALIVARY GLANDS. There are three pairs of large salivary glands whose ducts open into the mouth. There are also small salivary glands in the lips, cheeks, tongue and palate. The largest of the glands is the *parotid*, a wedge-shaped structure lying between the mastoid process behind and the angle of the mandible in front. It extends upwards as far as the temporomandibular joint, downwards over the sternocleidomastoid muscle and forwards over the masseter muscle (Fig.81). The facial nerve passes forwards and the external carotid artery upwards through it. Its duct passes forwards across the masseter muscle and then turns inwards. It pierces the buccinator muscle and opens into the mouth inside the cheek opposite the upper second molar tooth.

The *submandibular salivary gland* lies in the upper part of the neck under the skin, below the floor of the mouth and deep to the body of the mandible (Fig.81). It is about 2 to 3 cm in width, depth and

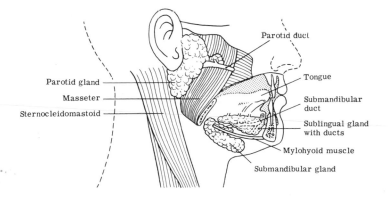

Fig.81. Salivary glands.

height. Its duct winds round the posterior edge of the muscle forming the floor and then medially in the fold between the tongue and the floor of the mouth. The duct opens into the mouth at the side of the frenulum (Fig.80b).

The *sublingual salivary gland* is much smaller, about the size and shape of a shelled almond, and lies in the fold between the floor of the mouth and the tongue. It has a number of ducts which open on to the summit of the fold (Figs.80b, 81).

The glands consist of a number of lobes and lobules each consisting of a branching duct at the end of which are somewhat dilated structures called alveoli (Fig.6c). These alveoli consist of either mucous or serous cells, that is, they produce either a thick mucous secretion or a thin watery secretion. The parotid gland consists entirely of serous alveoli. The other two are mixed.

The alveoli produce a secretion called *saliva* which passes into the mouth by means of the ducts. Saliva consists almost entirely of water (99·5 per cent) with some mucin. It contains an enzyme called *ptyalin*. It also contains some salts. The functions of saliva are to moisten food so that it can be made into a lump (*bolus*) for swallowing and also to lubricate its surface with mucus. In addition the ptyalin breaks down cooked starches into simpler sugars. The sense of taste does not function unless substances are dissolved, and saliva provides the fluid for this purpose. Saliva also moistens and lubricates the lining of the mouth, pharynx and oesophagus, and in this way helps articulation. Speech is difficult with a dry mouth. About 1,000 to 1,500 ml of saliva are produced in 24 hours. Excessive loss of water from the body, by whatever means, results in a diminution of the quantity of saliva secreted. This produces dryness of the mouth and a feeling of thirst. Thus reduction in salivary secretion can act as a warning mechanism that water is required by the body.

THE TONGUE. This is a muscular organ which lies partly in the floor of the mouth and partly in the lower part of the pharynx. It has a free anterior part and a more fixed posterior part. The upper surface is called the *dorsum* and has a V-shaped groove called the *sulcus terminalis* (Fig.80c). The apex of the V is posterior. Normally the upper surface of the anterior part is horizontal and that of the back part is vertical.

The mucous membrane on the upper surface of the front part of the tongue is thicker than that of the inferior surface. Both are covered by stratified squamous epithelium but the superior has *papillae* (projections) and the inferior is smooth. The *frenulum* is a vertical fold of mucous membrane in the midline between the tongue and the floor of the mouth. The mucous membrane over the base of the tongue is somewhat irregular due to the underlying *lingual tonsil*, consisting of lymphoid tissue. Behind the tongue is the epiglottis (Fig.80c).

There are three types of papillae on the tongue. The *vallate papillae* are about 1 to 2 mm in diameter and form a row on each side just in front of and parallel to the sulcus terminalis (Fig.80c). Each papilla is surrounded by a groove outside which there is an elevation of mucous membrane. In the walls of the groove are taste buds. *Fungiform papillae* are found chiefly on the sides and tip of the tongue and are smaller than the vallate (Fig.80c,d). They are circular but have no groove round them. These papillae also have taste buds. *Filiform papillae* are conical in shape and are found over the dorsum of the front part of the tongue (Fig.80c,d). These papillae have no taste buds. Furring of the tongue in illness is due to the surface cells of the filiform papillae accumulating on the tongue.

Most of the tongue consists of striated muscle, divided into *intrinsic* and *extrinsic muscles*. The intrinsic lie entirely in the tongue and are found mainly in its free part. These muscles are important in the first stage of swallowing, when the food is placed between the dorsum of the tongue and the hard palate and forced backwards into the oropharynx by pressure of the free part of the tongue on the hard palate.

The extrinsic muscles of the tongue connect the tongue to the base of the skull above and behind, to the hyoid bone below, to the jaw below and in front and to the soft palate above. The tongue can be raised, depressed, protruded and retracted by these muscles.

The tongue is used in speech, chewing and swallowing. It is also the organ of taste. The end organs of taste are the *taste buds*, which are found in the walls of the vallate papillae and in the fungiform papillae. All taste buds are structurally alike but there are four qualities of taste—bitter, sweet, sour and salt. Each of these is specially related to different parts of the tongue—sweet and salt to the tip, bitter to the back, and sour to the sides.

THE TEETH. There are two sets of teeth in man, *deciduous* (*milk*) and *permanent*. All teeth have a similar basic structure, a *crown* projecting above the gum, a *root* (or roots) in the jaw and a narrowed part, the *neck*, between the crown and the root (Fig.82a,b,c). The

Table 3

	Molars	Canine	Deciduous teeth Incisors	Incisors	Canine	Molars	
Upper jaw	2	1	2	2	1	2	
Lower jaw	2	1	2	2	1	2	

	Molars	Premolars	Canine	Permanent teeth Incisors	Incisors	Canine	Premolars	Molars
Upper jaw	3	2	1	2	2	1	2	3
Lower jaw	3	2	1	2	2	1	2	3

bulk of the tooth consists of *dentine*. Over the dentine of the crown is the *enamel* and round the dentine of the root is the *cement*. The *periodontal ligament* attaches the root of the tooth to its socket. There is a small opening in the apex of the root through which vessels and nerves pass to and from the *pulp*. Both enamel and dentine are very hard because of their high mineral content.

There are twenty deciduous teeth, five in each quadrant. There are, from the midline in front passing laterally and backwards, two

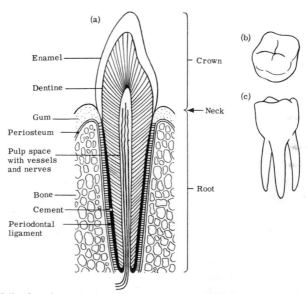

Fig.82. (a) Section through typical tooth, (b) Upper surface of crown of molar, (c) Upper molar with three roots.

incisors, one canine and two molars (Table 3). The first to appear
are the lower central incisors at about six months. All twenty teeth
are usually present by the end of the second year.

There are thirty-two permanent teeth, eight in each quadrant.
There are, from the midline in front and going backwards, two
incisors, one canine, two premolars and three molars (Table 3).
The first to appear is the first molar at about the sixth year. This
appears behind the second deciduous molar. The rest of the permanent teeth appear between the seventh and thirteenth years except
for the third molars which appear between the seventeenth and
twenty-fifth year and are known as the *wisdom teeth*. Sometimes the
third molars do not erupt. The permanent incisors appear after the
first molars, the subsequent order of appearance is premolars, canine
and molars.

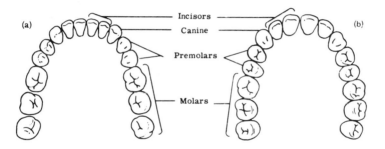

Fig.83. (a) Lower jaw teeth, (b) Upper jaw teeth.

The general form of the different teeth is as follows. The crowns
of the incisors have a horizontal, bevelled cutting edge and a single
root. The canines have a conical crown, and a long single root.
The premolars (*bicuspids*) have two cusps on the crown and a single
root (the first upper premolar usually has a double root). The molars
have three, four or five cusps on the crown. The upper molars have
three roots and the lower molars two. In general the incisors and
canine are for cutting and the premolars and molars for masticating.

The alveolar arch formed by the two maxillae has a different
curvature from that of the mandibular arch (Fig.83a,b). The upper
arch is wider in front and narrower behind than the lower. The
result is that when the jaws are together the front upper teeth overlap and lie in front of the front lower teeth. The outer half of the

upper molars overhangs that of the lower molars because the upper molars slope down and out and the lower molars up and in.

The teeth are used for cutting and masticating food.

THE TONSILS. There is a ring of lymphoid tissue round the upper parts of the respiratory and alimentary tracts. The ring is not continuous and its different parts have already been referred to. There is lymphoid tissue on the posterior wall of the nasopharynx (*pharyngeal tonsil*) and on the back of the upper surface of the tongue (*lingual tonsil*). On each side of the fauces, the region between the mouth and the oropharynx, there is the *palatine tonsil*, often referred to as the tonsil. All these tonsils are larger in a child than in an adult.

Their main function is protective, but clinically they are liable to become infected and enlarged. The pharyngeal tonsil, often referred to as the adenoid, can enlarge forwards and laterally. If it enlarges forwards it can obstruct the back of the nasal cavity so that the child of necessity becomes a mouth-breather. Lateral enlargement can affect the pharyngeal end of the Eustachian tube. The palatine tonsils are particularly liable to infection in childhood and this results in acute and chronic tonsillitis. Sometimes these conditions lead to disease elsewhere in the body. The lingual tonsil does not appear to be liable to infection as in the same way as the other tonsils.

The pharynx and oesophagus

The pharynx has already been described with the respiratory system. It extends from the base of the skull to the level of the sixth cervical vertebra where it is continuous with the oesophagus which is about 25 cm long.

The oesophagus is a tubular structure passing down through the lower part of the neck and the whole of the thorax. It pierces the diaphragm and ends by joining the right border of the stomach at its upper end (Fig.84). Its abdominal part is about 3 cm long. It lies more or less in the midline in front of the vertebral column but deviates somewhat to the left before it passes through the diaphragm. The oesophagus lies behind the trachea in the neck and upper part of the thorax and lower down behind the heart. In the thorax the arch of the aorta and the left bronchus lie to the left and in front of it, and the thoracic aorta is behind at its lower end above the diaphragm (Fig.84).

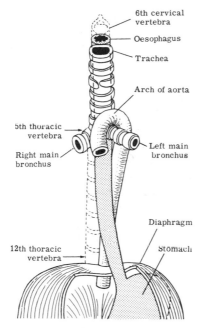

Fig.84. Oesophagus.

The oesophagus is narrowed, (a) at its beginning just below the pharynx, (b) where the left bronchus crosses it, (c) as it passes through the diaphragm. These are important sites because cancer of the oesophagus tends to occur more often in these positions.

Structurally the oesophagus consists of a lining of non-keratinized stratified squamous epithelium, a submucous layer containing mucous glands opening into the lumen, and inner circular and outer longitudinal muscular layers. The muscle is striated above and smooth below. Since the epithelium is stratified, solid food can remove only the outer layer of epithelium. The mucous glands lubricate the lining, and the striated muscle moves the food quickly in the upper part of the oesophagus.

Food when swallowed passes down the oesophagus due to a wave of peristalsis, that is a wave of contraction preceded by a wave of relaxation. The food is prevented from passing upwards from the stomach into the oesophagus by various mechanisms. It is thought that the diaphragm may act as a sphincter, the mucous membrane at the opening of the oesophagus into the stomach may act as a valve and the angle of union between the oesophagus and stomach is such as to prevent regurgitation of the food. There is no evidence of an anatomical sphincter at the lower end of the oesophagus. There is a condition called *achalasia* in which the lower end of the oesophagus goes into spasm and prevents the entry of food into the stomach. The food accumulates in the oesophagus and either the patient is sick or the spasm passes off and the food enters the stomach.

SWALLOWING (DEGLUTITION). Many references have already been made to this function of the mouth and pharynx. Swallowing may be divided into two stages, (*a*) the oral and (*b*) the pharyngeal.

    a. After the food is chewed by the teeth, a bolus is formed. Usually the lips are kept closed and the cheeks contract as the anterior part of the tongue is pressed on to the hard palate from in front backwards. This involves elevating the floor of the mouth, and an upward and backward movement of the tongue. When fluid is swallowed the tip and sides of the tongue are pressed on to the hard palate so that a trough is formed.

    b. The bolus (or fluid) passes through the isthmus of the fauces and enters the oropharynx. Stimulation of the fauces or walls of the oropharynx produces a series of reflex muscle contractions controlled by a deglutition centre in the medulla oblongata, the *swallowing reflex*. As a result, the food is prevented from going upwards into the nasopharynx or back into the mouth. The nasopharynx is shut off by the upward and backward movement of the soft palate. The mouth is shut off by the approximation of the folds or pillars of the fauces and the back of the tongue maintaining contact with the palate.

The bolus now passes down the pharynx into the oesophagus. It is prevented from entering the larynx by its elevation which brings the opening of the larynx close to the epiglottis. The entrance to the larynx is closed by the approximation of its edges in which there is muscle. At the same time the vocal folds come together. It is said that the vestibular folds also come together. There still is some argument as to whether the epiglottis is pulled down like a lid over the laryngeal opening. It should be added that food, especially fluid, may pass downwards along the sides of the laryngeal opening so that the possibility of food entering the larynx is very much reduced.

    Finally respiration is inhibited during swallowing and if any food or fluid gets into the larynx the cough reflex occurs and an attempt is made to eject the food from the larynx.

The abdominal cavity

The rest of the alimentary system lies in the abdominal and pelvic cavities. The walls of these cavities consist of bones and muscles. The lumbar and sacral vertebrae form part of the posterior wall and the bony pelvis forms most of the lower part of the cavities. Part of the back wall and side walls are muscular, and the anterior wall is wholly muscular. The roof is formed by the diaphragm and, because of its dome-shaped structure, the abdominal cavity extends upwards inside the lower ribs, below the diaphragm. The floor of the cavity is muscular and is called the *pelvic diaphragm.*

The abdomen is divided into nine regions by two vertical and two horizontal lines. Since some of these regions are frequently mentioned by name it is necessary to know which region is referred to. Fig.85 indicates the names and sites of the main regions.

THE PERITONEUM. The abdominal cavity is lined by *peritoneum,* and all the abdominal organs are outside the peritoneum, between it and the abdominal wall (Fig.86a,b). The organs project into the cavity and therefore have a covering of peritoneum. The peritoneum lining the adbominal cavity is called *parietal* and that covering an

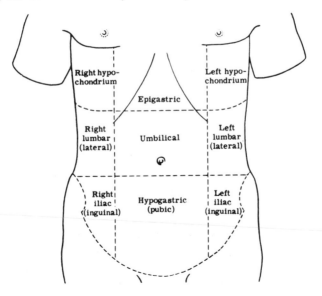

Fig.85. Regions of abdomen.

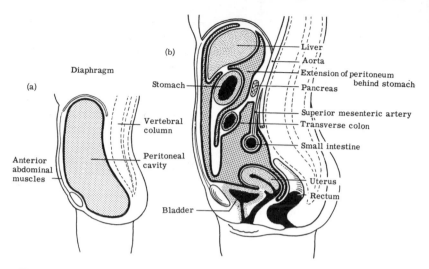

Fig.86. (a) Theoretical peritoneum and peritoneal cavity, (b) Actual peritoneum and peritoneal cavity in female.

organ is called *visceral*. Some of the organs are close to the abdominal wall and others are suspended from the wall by a double fold of peritoneum called a *mesentery*, in which the blood vessels of the organ are found (Fig.87). During development the arrangement of the peritoneum becomes very complicated and it is difficult to simplify this. One of the complications is the extension of the back of the peritoneum *behind* the stomach to the left as far as the spleen (Fig.86b). This leftward extension passes upwards behind the liver and downwards into the double fold of peritoneum, the *greater omentum*, hanging from the stomach. This can be seen in a longi-

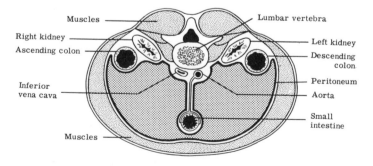

Fig.87. Transverse section through middle of abdominal cavity.

tudinal section of the abdominal cavity as is shown in Fig.85b. The importance of the peritoneum is that, like the other serous membranes, the pleura and the serous pericardium, it enables organs to move easily on each other. The peritoneum has another function in that it is protective. It contains large numbers of phagocytic cells, cells which can ingest harmful organisms and an infected organ, for example, the appendix in appendicitis, may be surrounded by the greater omentum which hangs freely in the abdominal cavity (Fig.88).

It will be appreciated that operations through the anterior abdominal wall take place inside the peritoneal cavity, but some operations are performed outside the peritoneum, for example, a kidney is usually removed from behind so that the peritoneal cavity is not opened. The bladder, when it fills, enlarges upwards and strips off the peritoneum from the anterior abdominal wall. It is not unknown for a surgeon to cut into the bladder if the patient has been kept waiting a long time in the anaesthetic room.

Hernia has been defined as a protrusion of the lining membrane of a body cavity together with some part of the organ or organs inside the cavity. Peritoneal hernias are by far the commonest and of these the hernia through the inguinal canal is the most frequent, because in the course of development everybody has a protrusion of peritoneum through this canal (Fig.48). In most people the protrusion disappears. If it persists it is likely that some organ, usually a loop of small intestine, will pass into the protrusion and produce a visible hernia. There is also a canal medial to the

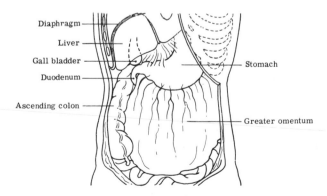

Fig.88. Appearance of abdominal cavity after removal of anterior abdominal wall.

femoral vessels deep to the inguinal ligament. A femoral hernia can develop in this canal (Fig. 48). Other common types of hernia are those seen alongside the umbilicus (*umbilical hernia*) and after an abdominal operation (*incisional hernia*). A hernia upwards into the thorax through the opening in the diaphragm for the oesophagus can also occur (*diaphragmatic hernia*).

THE GENERAL STRUCTURE OF THE ABDOMINAL PART OF THE ALIMENTARY TRACT. The alimentary tract may be regarded as a tube extending from the mouth to the anus. The tube is very much changed in structure in the region of the mouth and considerably modified in the pharynx. However, it shows relatively minor variations from the oesophagus to the anus. Basically the tube consists of a lining (mucous membrane) covered by columnar epithelium, a submucous coat outside that, then two layers of smooth muscle (an inner circular and an outer longitudinal) and finally a serous coat which is peritoneum (Fig.89a). Some of the modifications of this basic structure have already been seen in the oesophagus—the epithelium is stratified squamous, the muscle is striated in its upper third and there is no serous covering except in its abdominal part.

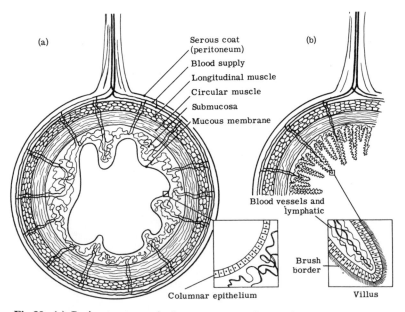

Fig.89. (a) Basic structure of alimentary tract (inset-columnar epithelium of lining), (b) Structure of small intestine (inset-structure of villus).

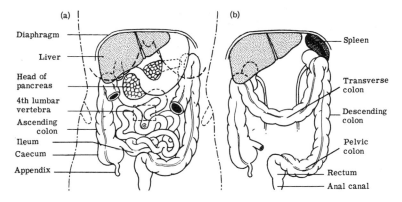

Fig.90. (a) General arrangement of abdominal viscera (liver has been pulled upwards), (b) Position of large intestine.

The modifications of the different parts of the tube will be mentioned.

The general arrangement of the abdominal organs is shown in Fig. 90a,b.

THE STOMACH. The *stomach* is approximately J-shaped and lies mainly to the left of the midline below the left half of the diaphragm. The oesophagus enters the upper part of its right border at the *cardiac orifice*. The stomach is continuous with the first part of the *small intestine* (*duodenum*) to the right of the first lumbar vertebra at the *pyloric orifice*. The right border of the stomach is called the

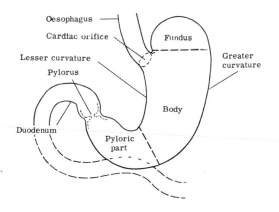

Fig.91. Stomach.

lesser curvature and the left the *greater curvature*. The part above the entrance of the oesophagus is called the *fundus*, the part next to the pyloric orifice, the *pyloric part*, and the part between, the *body* (Fig.91). In front of the stomach is the diaphragm, the left lobe of the liver and the anterior abdominal wall. Behind the stomach from right to left are the pancreas, left kidney and spleen.

The stomach has a lining of mucous, columnar epithelium with many glands extending down into the underlying loose connective tissue. The muscle of the stomach is arranged in three layers, an outer longitudinal, a middle circular and an inner oblique. The additional layer of muscle is probably related to the churning of the food in the stomach. The circular layer is very much thickened round the pyloric orifice and forms the *pyloric sphincter*. The stomach is covered by peritoneum.

The stomach is the most dilated part of the alimentary tract. It acts as a storage chamber and food remains in it for two to four hours. The food is mixed with the secretion of the glands and partly digested. This product is called *chyme*. The digestion of proteins begins in the stomach and alcohol can be absorbed by the stomach to some extent.

The chyme enters the duodenum in small quantities due to contraction of the gastric muscle and relaxation of the pyloric sphincter. Distension of the duodenum reflexly inhibits gastric movements and gastric motility decreases. As the duodenum empties this inhibitory action ceases and the stomach again contracts, the pyloric sphincter relaxes and chyme enters the duodenum.

THE SMALL INTESTINE. The *small intestine* is about five metres long extending from the pyloric orifice to the *ileocaecal orifice* where it joins the *large intestine* in the lower right part of the abdominal cavity (Fig.90a). The small intestine is coiled and lies in the middle of the cavity behind the anterior abdominal wall. It is suspended from the posterior abdominal wall by a double fold of peritoneum—the *mesentery*.

The first part, the *duodenum*, is about 25 cm long and is shaped like a letter C curving round the right end of the *pancreas* which lies in front of the second lumbar vertebra (Fig.92). The duodenum begins at the pyloric sphincter to the right of the first lumbar vertebra and ends at the *duodenojejunal flexure* to the left of the second lumbar vertebra. The bile duct and pancreatic ducts open into the back wall of the second part of the duodenum by a common orifice.

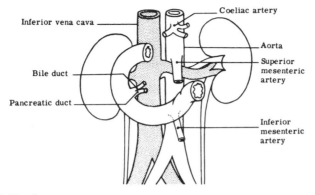

Fig.92. Duodenum.

By means of these ducts the *bile* (from the liver) and the pancreatic secretions containing digestive enzymes enter the small intestine.

The next two-fifths of the small intestine are called the *jejunum* and the final three-fifths, the *ileum*. The terminal part of the ileum passes upwards out of the pelvis to join the large intestine at the ileocaecal orifice.

The lining of the small intestine has a greatly increased surface area because it is elevated into circular folds and on the folds are projections called *villi* (Fig.89b). The mucous membrane, covered by mucous columnar epithelium, also extends into the underlying tissue as glands. There are two muscle coats, an outer longitudinal and an inner circular, and except for the duodenum the small intestine is almost completely surrounded by peritoneum.

The functions of the small intestine are to continue the digestion of the food which leaves the stomach and absorb the broken-down products of digestion. The almost fluid residue in the terminal part of the ileum passes into the large intestine through the ileocaecal orifice. The digestion of proteins, carbohydrates and fats is carried out by means of enzymes secreted in the pancreatic juice and the intestinal juice. These secretions are produced as the result of nervous and hormonal influences acting on the pancreas and small intestine. Bile from the liver assists in the digestion of fats and is stored in the gall bladder which empties when a fatty meal enters the duodenum.

The end-products of the digestion of the proteins, carbohydrates and fats are amino acids, monosaccharides (mainly glucose) and glycerol and fatty acids. These are absorbed by the lining of the small intestine. Both digestion and absorption are assisted by the movements of its wall. These movements are referred to as

a. *segmented* in which adjacent areas of the wall contract and relax so that the contents are pushed to and fro, and

b. *peristaltic* in which the food is pushed along the lumen.

At the end of the ileum the ileocaecal valve opens and closes to allow small quantities of what is left in the ileum to enter the large intestine. This includes mineral salts and vitamins.

Absorption of the broken-down products of digestion takes place through the villi into their capillaries. The blood in the capillaries passes to the liver through the *portal system of veins*. Some of the fatty acids and fat-soluble vitamins are absorbed into a lymphatic vessel in the middle of the villus and consequently go to the thoracic duct. They then enter the innominate vein and so go to the systemic venous circulation without first going through the liver.

THE LARGE INTESTINE. The *large intestine* is about 1·5 metres long and is divided into several parts (Fig.90b). Passing upwards on the right side of the abdominal cavity is the *ascending colon*. Below the region where the ileum joins the large intestine is the *caecum* which is about 6 cm long and 7·5 cm wide. Attached to the caecum below and to the left is the (*vermiform*) *appendix* which is about 9 cm long and about 1·5 cm wide. The appendix has a large amount of lymphoid tissue in its submucosal layer. At the upper end of the ascending colon behind the liver and in front of the right kidney, the large intestine turns to the left at the *right colic flexure* and becomes the *transverse colon* which is suspended from the posterior abdominal wall. The transverse colon turns downwards at the *left colic flexure* next to the spleen in front of the left kidney and becomes the *descending colon* which lies on the left side of the abdominal cavity. Lower down the large intestine forms a loop (*pelvic colon*) which is suspended from the left side of the posterior abdominal wall and hangs into the pelvis. The pelvic colon passes towards the midline into the hollow of the sacrum. The *rectum* lies in the midline, is about 12 cm long, and passes downwards. It bends backwards at a right angle and becomes the *anal canal*, the terminal 3 cm of the alimentary tract. The anal canal opens on to the exterior at the *anus*.

The large intestine has a lining of mucous epithelium with a very large number of mucous cells. The inner circular layer of unstriped muscle is complete but the outer longitudinal is in the form of three longitudinal bands called the *taeniae coli*. These are shorter than the true length of the large intestine and account for its puckered appearance. The longitudinal muscle is more or less complete in the appendix and the rectum.

The rectum and anal canal have sphincters, thickenings of circular muscle. The sphincter round the terminal part of the anal canal consists of striated muscle. The peritoneum more or less surrounds the appendix, caecum, transverse colon and pelvic colon but it covers only the anterior surface of the ascending colon, descending colon and rectum. The anal canal is below the level of the peritoneum (Fig.86b).

The fluid material which enters the caecum and ascending colon is semi-solid by the time it reaches the exterior as *faeces*. This is due to the absorption of water which takes place in the large intestine. In addition mineral salts are absorbed. Glucose and drugs, if administered per rectum, are absorbed. The mucous glands produce mucus which acts as a lubricant.

Faeces consist mainly of water (75 per cent) and the indigestible remains of the food. This is largely cellulose from fruit and vegetables. Its colour is due to bile pigments and its odour due to gases (indole and skatole) produced by the action of bacteria of which there are large numbers in the faeces. It takes about $4\frac{1}{2}$ hours for the remains of a meal to reach the caecum and about 16 hours to reach the rectum. Defaecation, which in the adult is a voluntary act, is due to contraction of the descending colon, pelvic colon and rectum assisted by contraction of the abdominal muscles, the fixing of the diaphragm in inspiration and the relaxation of the sphincters. The desire to defaecate is stimulated by distension of the rectum by faeces.

The blood supply of the alimentary tract, from the stomach to the rectum, is from three arteries which arise from the abdominal aorta (Fig.92). The veins corresponding to these arteries go to the liver via the portal vein. The muscle of the alimentary tract is supplied by autonomic nerves, both sympathetic and parasympathetic.

THE LIVER. This organ is the largest gland in the body and weighs about 1·5 kg. It lies mainly in the upper right part of the abdominal cavity below the right half of the diaphragm within the right half of the thoracic cage and extends to the left of the midline for about 3 cm. It is divided into a large right and small left lobe by the mid-

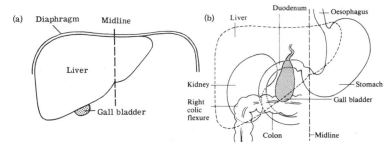

Fig.93. (a) Liver, (b) Organs related to visceral surface of liver.

line. Most of the surface of the liver is in contact with the diaphragm and it has an inferoposterior (or visceral) surface which faces downwards and backwards (Fig.93a,b).

In the middle of the visceral surface is the *porta hepatis* into which go the hepatic artery and portal vein and out of which comes the hepatic duct (Fig.93b). The inferior vena cava is more or less lodged in the upper part of this surface to the right of the midline and the hepatic veins, draining the blood from the liver, enter the inferior vena cava. The abdominal oesophagus and the stomach lie behind the left lobe and the right colic flexure, right kidney and duodenum behind the right lobe.

The *gall bladder* lies on the lower part of the visceral surface to the right of the midline. If followed upwards the gall bladder becomes continuous with the *cystic duct* which joins the hepatic duct to form the bile duct (Fig.94). This duct passes downwards and to the right behind the first part of the duodenum and joins

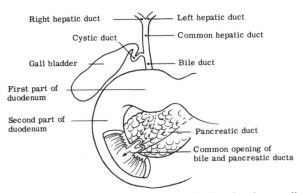

Fig.94. Extrahepatic bile duct system (the gall bladder has been pulled to the right).

the pancreatic duct in the wall of the second part of the duodenum into which they have a common opening.

The lobes of the liver are subdivided into lobules which consist of hexagonally arranged columns of cells at the angles of which are found branches of the hepatic artery, portal vein and bile duct (Fig. 95a,b). In the centre of a lobule is a vein which ultimately joins the hepatic veins. The blood passes from the branches of the portal vein and hepatic artery between the liver cells in wide channels, called *sinusoids*, to the veins in the centre of the lobules (Table 4).

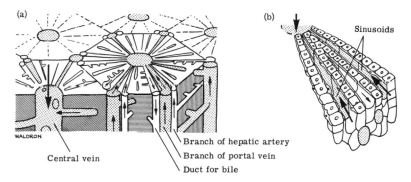

Central vein
Branch of hepatic artery
Branch of portal vein
Duct for bile

Fig.95. (a) Lobules of liver with circulation of blood and flow of bile, (b) Arrangement of liver cells in relation to flow of blood and bile.

Table 4

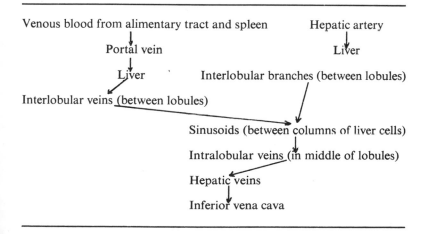

One of the functions of the liver is to produce bile which passes from the cells towards the bile capillaries between the lobules and ultimately reaches the hepatic duct. The bile is stored in the gall bladder (capacity 30 to 50 ml) and there it is concentrated ten times by the absorption of water. When fat enters the small intestine the bile is expelled by contractions of the muscular wall of the gall bladder into the duodenum along the cystic and bile ducts. Bile contains bile salts which are involved in the digestion of fats and absorption of fatty acids. It also contains bile pigments which are derived from broken-down haemoglobin. Bile pigments are really waste products and appear in the faeces and, after absorption and modification, in the urine.

THE PANCREAS. The pancreas is a soft lobulated gland about 15 cm long lying on the posterior abdominal wall. It extends from the concavity formed by the duodenum across the left kidney to the spleen. To the right it is rounded but towards the left it is considerably tapered. It lies behind the stomach (Fig.90a). Its duct passes from left to right in the substance of the gland and at its termination joins the bile duct to open into the second part of the duodenum.

The gland consists of lobules which consist of a main duct, its subdivisions and terminal alveoli. These alveoli produce an external secretion, *pancreatic secretion*. It consists mainly of water and is markedly alkaline due to the presence of a large amount of bicarbonate. It contains enzymes which act on proteins, fats and carbohydrates. About 500 to 800 ml of pancreatic juice are secreted daily.

Between the alveoli are collections of cells called the *islets of Langerhans*. These produce a hormone called *insulin* which passes directly into the blood. Insulin plays a very important part in the metabolism of glucose. It is essential for the formation of glycogen from glucose and the storage of glycogen in the liver and in muscle. Insulin also acts on all cells which use glucose in their metabolism. In the absence of insulin (as in *diabetes mellitus*) glycogen deposition in the muscles and liver is decreased and the glucose in the blood is raised. Instead of being re-absorbed in the renal tubules glucose is excreted in the urine.

THE SPLEEN. This organ is included because it lies in the abdominal cavity and because its blood supply is from the same artery which supplies the stomach and liver and parts of the pancreas and duo-

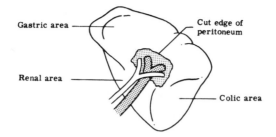

Gastric area

Cut edge of peritoneum

Renal area

Colic area

Fig.96. Visceral (medial) surface of spleen.

denum, and its venous blood goes to the liver in the portal vein. It plays no part in the digestion or absorption of food. The spleen is dark purplish in colour and is about 12 cm long, 6 cm wide and 5 cm thick but it is variable in size. It lies in the upper posterior part of the left side of the abdominal cavity. Its right surface is in contact with the stomach, left kidney and left colic flexure and its left surface is in contact with the diaphragm (Fig.96). Entering and leaving the deep surface are the splenic artery and splenic vein respectively. The spleen consists largely of lymphoid tissue.

In view of its structure the spleen must be a source of lymphocytes. The spleen is also important in the defence mechanisms of the body and is associated with the production of immune bodies. Degenerating blood corpuscles, lymphocytes and blood platelets are destroyed in the spleen. In the fetus the spleen forms red blood corpuscles but after birth this function ceases and is resumed only in an emergency. The spleen may act as a reservoir of red blood corpuscles.

9

The urinary system

This consists of two kidneys, each with a ureter opening into the bladder. The bladder opens on to the exterior by means of the urethra (Fig.97a,b). In the male this is about 18 cm long and passes through the prostate and penis; in the female it is about 5 cm long and opens in front of the opening of the vagina (Fig.97b). It is

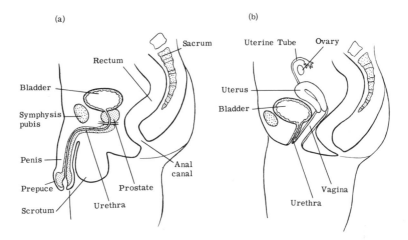

(a) (b)

Fig.97. (a) Sagittal section through male pelvis, (b) Sagittal section through female pelvis.

common for the urinary system to be associated with the genital system (*urogenital system*) and from an embryological point of view this is justifiable. It is probably more convenient to separate the two systems.

154

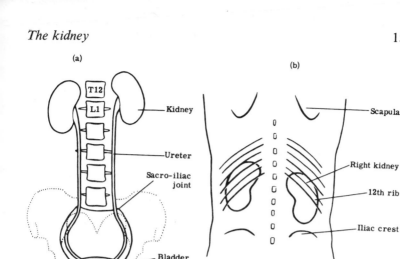

Fig.98. (a) Position of urinary organs relative to vertebral column and pelvis, (b) Position of kidney as seen from behind.

The kidney

Each *kidney* lies on either side of the vertebral column on the muscles of the posterior abdominal wall at the level of the twelfth thoracic and first and second lumbar vertebrae (Fig.98a,b). It has a characteristic shape and on its medial concave border is the *hilum* into which enters the renal artery and from which emerge the renal vein and ureter. An average kidney is about 10 cm long, 6 cm wide and

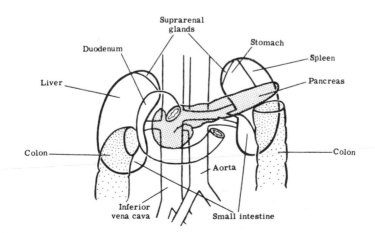

Fig.99. Organs related to kidneys on posterior abdominal wall.

3 cm thick. The right kidney is slightly lower than the left, probably because of the liver. The small and large intestine lie in front of the lower half of both kidneys. The suprarenal gland is related to the upper part. The liver and duodenum are in front of the right kidney, and the stomach, spleen and pancreas are in front of the left kidney (Fig.99).

If a kidney is divided longitudinally from its concave border to its convex border, the cut surface can be seen to be divided into an outer zone, the *cortex*, and an inner zone, the *medulla* (Fig.100a). In the medulla are paler, conical masses called the *renal pyramids* and between the pyramids the cortex continues into the medulla as the *renal columns*. The ureter is seen to expand into the *pelvis* of the kidney. The pelvis is formed by the union of two of three subdivisions called *major calyces* which in turn are formed by the union of several

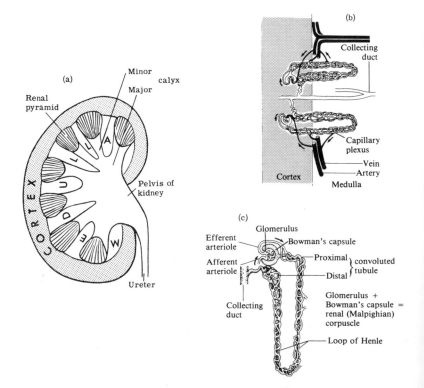

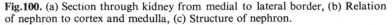

Fig.100. (a) Section through kidney from medial to lateral border, (b) Relation of nephron to cortex and medulla, (c) Structure of nephron.

minor calyces. The apices of the conical pyramids project into the minor calyces.

THE NEPHRON. The unit of the kidney is the *nephron*. This is essentially a tubular structure which ends by opening into a *collecting tubule* (Fig.100b,c). The beginning of a nephron is enlarged and forms a *renal (Malpighian) corpuscle* which consists of a tuft of blood vessels (*glomerulus*) invaginated into the blind end of the nephron (*Bowman's capsule*). All the renal corpuscles, of which there are about two to three million, are in the cortex. Bowman's capsule leads into a convoluted part of the tubule, also in the cortex, and this loops into a pyramid of the medulla (*loop of Henle*, with a descending and ascending limb). The ascending limb enters the cortex and becomes convoluted before entering a collecting tubule which opens into a duct leading to a minor calyx at the apex of a pyramid.

THE FUNCTIONS OF THE KIDNEY. The functions of the kidney are:

    a. to maintain water balance,

    b. to maintain acid–base equilibrium,

    c. to excrete waste products such as urea.

These functions are carried out by filtering large quantities of protein-free fluid from the blood (about 200 l per 24 hours). The blood is brought to the nephron by an afferent vessel and taken away by an efferent vessel. Fluid containing glucose, crystalloids (mainly sodium chloride) and waste products pass through Bowman's capsule mainly due to the blood pressure being greater than the osmotic pressure of the plasma proteins. Glucose is normally completely re-absorbed in the proximal tubule and most of the water and sodium chloride is also re-absorbed in this part of the nephron. While the remaining filtrates pass along the loop of Henle sodium chloride is absorbed and under the influence of the antidiuretic hormone produced by the hypophysis cerebri (pituitary gland) more water is absorbed through the walls of the terminal part of the tubule and the collecting ducts. It was thought that the re-entry of water into the blood vessels took place in the loop of Henle and was due to the osmotic pressure of the plasma proteins being greater than the pressure of the fluid in the tubule. This is not now believed to be the case. Some of the urea is re-absorbed.

Urine is the end-product of the processes of filtration and re-absorption in the kidney. About 1500 ml are excreted daily and urine consists of about 95 per cent water, 2 per cent urea and 2 per cent sodium chloride. The amount of urine varies with the fluid intake and the water lost by other means such as sweating. Similarly the amount of sodium chloride excreted varies with the amount lost by sweating. Normally the urine is slightly acid but it can be made more or less acid or even alkaline in order to maintain the acid–base equilibrium of the body. Urea is the waste substance produced by the liver during de-amination of the excess amino acids in the body. The urine also contains other waste products of protein metabolism, for example, uric acid and creatinine.

The ureter

The *ureter* is about 25 cm long and 0·5 cm wide and is a narrow muscular tube beginning at the pelvis of the kidney and running downwards on the posterior abdominal wall at the side of the vertebral column. It enters the pelvis at the sacro-iliac joint, runs forwards and joins the upper lateral angle of the triangular posterior surface of the bladder (Fig.98a). The function of the ureter is to convey the urine from the kidney to the bladder. This is associated with waves of peristalsis along the ureter occurring at intervals of about 20 seconds.

The bladder

The *bladder* is a hollow muscular organ lying in the pelvis behind the symphysis pubis in front of the rectum in the male and in front of the uterus and vagina in the female (Fig.97a,b). Its size varies with the amount of urine in it and it is usually emptied when it contains about 300 ml (about half a pint). As it fills it enlarges upwards behind the anterior abdominal wall outside the peritoneum.

The bladder has a fairly fixed, posterior, triangular wall (the *base*). Entering the upper lateral angles of the base are the ureters and leaving the lower angle in the midline is the urethra. In the male the prostate lies below the bladder and on each side on the back of the base are a ductus (vas) deferens and a seminal vesicle (Fig.103a). In the female the uterus tends to lie on top of the bladder when it is empty (Fig.97b).

The ureters pass obliquely through the wall of the bladder so that when the bladder contracts the ureters are closed and urine cannot pass up the ureters. There is a sphincter of smooth muscle round the beginning of the urethra.

The bladder is a reservoir for the urine and is capable of considerable distension. Its muscle is smooth and is innervated by autonomic nerves. The parasympathetic nerves from the sacral spinal nerves are motor to the muscle. Within certain limits the emptying of the bladder is an example of the voluntary control of involuntary muscle.

The urethra

The *urethra* extends from the bladder to the exterior. In the male it is about 18 cm long and passes through the prostate gland, then through a muscle called the *external urethral sphincter* and finally through the *corpus spongiosum* of the penis (Fig.97a). In the female the urethra is about 5 cm long. It passes downwards behind the symphysis pubis and is attached to the anterior wall of the vagina. It opens on to the exterior in front of the vaginal opening and behind the clitoris (Fig.108). The female urethra has an external sphincter near its termination.

Opening into the male urethra in its prostatic part are the two ejaculatory ducts and also the ducts of the prostatic glandular tissue (Fig.103).

The female urethra has a more or less consistent diameter and can be readily dilated to about 1 cm. The male urethra has a variable diameter and is narrowest at its external opening on the surface of the penis and where it passes through the external sphincter. It is widest in its prostatic part.

In both sexes the urethra consists of a lining of mucous membrane and a wall of connective tissue and smooth muscle. The internal sphincter at its upper end next to the bladder consists of smooth muscle. The external consists of striated muscle and after infancy is usually under voluntary control.

In micturition the muscle of the bladder contracts and the sphincters relax. The emptying of the bladder is assisted by contraction of the muscles of the anterior abdominal wall.

10

The genital system

The male genital organs

These include the testis and epididymis, the ductus deferens, the seminal vesicle and ejaculatory duct on each side, and the midline prostate and penis.

THE TESTIS AND EPIDIDYMIS. The *testis*, the male gonad, is an oval organ, about 5 cm long, 3 cm wide and 3 cm from front to back. It lies in a pouch of skin called the *scrotum*. Behind it lies the *epididymis*, shaped like a comma and attached to the back of the testis (Figs.101a, 102a). The testis and epididymis are covered by a double layer of serous membrane (the *tunica vaginalis*) derived from the peritoneum. The testis consists of a number of lobules (about 250) each containing two or three small convoluted tubules embedded in loose connective tissue. Each tubule if unravelled is about 75 cm

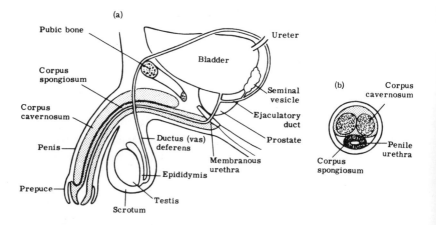

Fig.101. (a) Male genital organs, (b) Transverse section through penis.

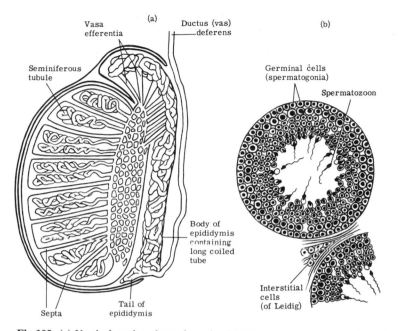

Fig.102. (a) Vertical section through testis, (b) Microscopic structure of seminiferous tubule.

long and about 0·2 mm in diameter. The *spermatozoa* are formed in the tubules. The tubules pass backwards and unite to form about twenty to thirty straight ducts which form a network of ducts passing upwards towards the upper end of the epididymis (Fig.102a).

The epididymis itself is a very long, narrow duct about 600 cm long and very much folded. The formation of the spermatozoa in the tubules of the testis involves a series of complicated changes which the cells on the periphery of the tubules undergo. These changes are associated with division of the cells and during one of these divisions the number of chromosomes is halved, from 46 to 23. As a result of the changes in shape a large round cell becomes a spermatozoon which in addition to other smaller structures consists mainly of a mass of nuclear material (chromosomes) and a long tail (Fig.102b). By means of this tail a spermatozoon can swim. It is said to be able to move about 2 to 3 mm per minute.

THE DUCTUS DEFERENS, SEMINAL VESICLE AND EJACULATORY DUCT. At the lower end of the epididymis the duct becomes

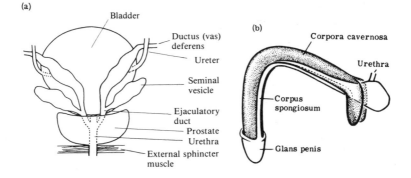

Fig.103. (a) Posterior surface of bladder showing relations of ductus deferens, seminal vesicle and ejaculatory duct, (b) Structure of penis.

larger and thicker and emerges as the *ductus* (*vas*) *deferens* which passes upwards towards the groin. The ductus which is about 50 cm long passes through the anterior abdominal wall just above the groin and enters the pelvis where it runs medially towards the back of the bladder. It then runs downwards on the back of the bladder where it has a diverticulum on its lateral side called the *seminal vesicle*, about 5 cm long (Fig.103a). Beyond the union with the seminal vesicle the ductus is called the *ejaculatory duct*. This duct passes downwards and forwards through the prostate and opens, one on each side, on to the prostatic urethra. The rest of the urethra is the common terminal passage for the spermatozoa and the urine. The ductus consists mainly of smooth muscle.

THE PROSTATE. The *prostate* is a glandular, muscular organ about 3 cm wide, 3 cm high and 2 cm from front to back. It lies below the bladder (Figs.97a, 101a). Through it pass the urethra and the ejaculatory ducts. The prostate produces a secretion containing enzymes but its exact functions are not known. Similarly the seminal vesicles produce a secretion which forms a considerable part of the seminal fluid.

THE PENIS. The *penis* consists of three longitudinal masses of vascular erectile tissue, the paired *corpora cavernosa* lying close together with the unpaired *corpus spongiosum* lying in the groove between them (Figs.101b, 103b). The corpus spongiosum forms the whole of the enlarged terminal part of the penis (the *glans*). The skin covering the penis extends beyond the proximal edge of the glans as the *prepuce* (foreskin) (Fig.101a). The urethra passes through the corpus spongiosum and the glans. Erection of the penis is due to the distension of the vascular spaces of the corpora.

THE FUNCTIONS OF THE TESTIS. The testis not only produces spermatozoa. It also secretes the male sex hormone *testosterone* which is formed by the *interstitial cells* in the connective tissue round the tubules. This hormone is produced at puberty under the influence of one of the hormones produced by the pituitary gland. It is responsible for what are known as the secondary sexual characteristics—the development of the testes and penis at puberty, the growth of hair on the face and elsewhere, the change in the voice due to the elongation of the vocal folds and certain aspects of the male emotional make-up.

The seminal fluid consists of the spermatozoa and the secretions produced by the epididymis and glandular structures, such as the prostate and seminal vesicles. About 3 ml of fluid are produced in each ejaculation and there are about 100 million spermatozoa per ml. The number, form and motility of the spermatozoa are important factors in male fertility. Less than 60 million per ml or many abnormal types or reduction of motility results in reduced fertility.

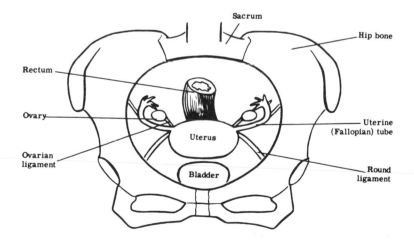

Fig.104. Internal female genital organs.

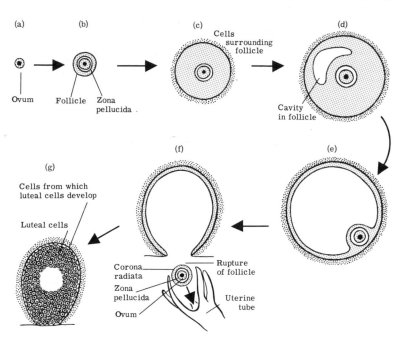

Fig.105. (a) Ovum in ovary, (b, c, d) Formation of ovarian (Graafian) follicle, (e) Ripe follicle, (f) Rupture of follicle with freeing of ovum, (g) Formation of corpus lutenum.

The female genital organs

The female genital organs include the ovary and uterine tube on each side and the midline uterus and vagina (Figs.97b, 104).

THE OVARY. The *ovary*, the female gonad, is almond-shaped and about 3 cm long, 1·5 cm wide and 1 cm thick. It is attached to the upper (posterior) layer of peritoneum passing from the uterus to the side wall of the pelvis (Fig.104). Its position varies but the ovary often lies on the lateral wall of the pelvis. The ovary consists of a framework of connective tissue cells in which are the *ovarian (Graafian) follicles*.

At first a follicle consists of a large central cell, surrounded by a single layer of smaller cells (Fig.105). The large central cell ultimately becomes the ovum (a process known as the *maturation of the follicle*), and the ovum is set free on the surface of the ovary. This release is called *ovulation*. In the process of maturation, brought about by a hormone of the pituitary gland, the follicle grows larger and the ovum is more or less surrounded by fluid within a thickened outer layer. The ovum also divides, usually twice. These divisions are unequal in that one large cell and one smaller cell are produced. During the first of these divisions the number of chromosomes in the ovum is reduced from 46 to 23.

Following ovulation the wall of the ovarian follicle collapses. Its cells increase in number and size and fill the follicle (Fig.105g). It is then called a *corpus luteum* because of the yellowish pigment in the cells. The development of the corpus luteum is controlled by another hormone produced by the pituitary gland.

The ovary also produces hormones called *oestradiol* and *progesterone*. The former, one of the oestrogens, is produced by the follicle and the latter by the corpus luteum. These hormones are responsible for the secondary sexual characteristics at puberty in the female—the enlargement of the uterus and vagina, the enlargement of the breasts, the menstrual cycle, the typical female distribution of fat and hair and psychological changes characteristic of the female. They are also important in the formation of the placenta (progesterone) and the changes in the breast during pregnancy (both hormones). Oestradiol is responsible for the repair and early development of the lining of the uterus after menstruation, and progesterone

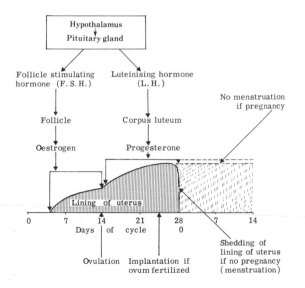

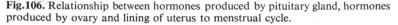

Fig.106. Relationship between hormones produced by pituitary gland, hormones produced by ovary and lining of uterus to menstrual cycle.

for its thickening and glandular enlargement in preparation for the embedding of the fertilized ovum (Fig.106). If the ovum is not fertilized the lining of the uterus is shed. This is associated with the degeneration of the corpus luteum.

THE UTERINE (FALLOPIAN) TUBE. The *uterine tube* is about 10 cm long and lies in the upper border of the broad ligament (Fig.107a). Its lateral end is irregular (fimbriated) and opens into the peritoneal cavity, and its medial end opens into the cavity of the uterus. It is lined by ciliated columnar epithelium and has a muscular wall. The ovum when shed at ovulation finds its way into the lumen of the tube. The cilia of the lining epithelium beat towards the uterus and cause a movement of fluid towards the uterus. This carries the ovum into the tube. It is also thought that the lateral end of the tube is more or less wrapped round the ovary so that on ovulation the ovum is shed into the lumen of the tube.

Fertilization, if it occurs, takes place in the lateral end of the tube. The spermatozoa swim through the cavity of the cervix and body of the uterus and into the lumen of the uterine tube.

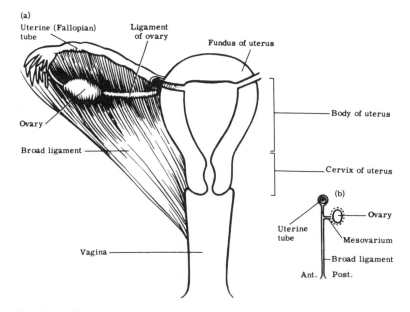

Fig.107. (a) Posterior view of female genital organs (the uterus and vagina are sectioned in the coronal plane), (b) Vertical section through broad ligament to show relation of this ligament to uterine tube and ovary.

THE UTERUS. The *uterus* is a hollow, muscular organ lying in the pelvic cavity in front of the rectum and behind the bladder. It is about 7·5 cm long, 5 cm wide and 2·5 cm from front to back. Its upper end is enlarged and its smaller lower end (*cervix*) projects into the vagina (Figs.97b, 107a). Usually it is bent forwards over the bladder. The wall is about 1 cm thick and the cavity is small. Opening into the upper part of the cavity on each side is the uterine tube, and the canal of the cervix opens below into the vagina.

The uterus is lined by columnar epithelium with many glands and the wall consists mainly of smooth muscle. The lining undergoes cyclical changes under the influence of the ovarian hormones. The lining of the uterus during the second half of the menstrual cycle thickens in order to receive a fertilized ovum. The changes in the lining (the *endometrium*) early on in the cycle consist of a replacement of the columnar epithelium and the re-formation of the glands. In the second half of the cycle, the glands become enlarged and swollen with secretion, and fluid appears between the cells in which the glands are embedded.

A fertilized ovum takes about four days to travel along the uterine tube and reach the cavity of the uterus. During this time it is dividing and developing and after about another four days in the cavity of the uterus it has developed sufficiently to be able to penetrate the wall of the uterus (*implantation*) at which stage the endometrium is ready to receive the embryo. This is about the 23rd day of a 28-day cycle. The corpus luteum of the ovary continues to function and under its influence the embryo continues to develop. The uterus enlarges to accommodate the growing fetus. At the end of pregnancy the uterine muscle, assisted by other muscles, contracts and expels the fetus, and placenta and membranes. After about six weeks, it returns approximately to its former size.

If pregnancy does not occur, the lining of the uterus is shed together with a variable quantity of blood. The deepest part of the lining, containing fragments of the glands, remains and from these the new endometrium is developed.

The position of the uterus in the pelvis is maintained by the connective tissue round the cervix and the muscles forming the floor of the pelvis (*pelvic diaphragm*). During childbirth these tissues and muscles are stretched and may be torn. As a result the uterus descends through the pelvis, a condition known as *prolapse*. Prolapse is often associated with disturbances of micturition. It is treated by an operation involving a repair of the stretched and torn structures in the floor of the pelvis.

The cervix of the uterus, although showing cyclical changes during the menstrual cycle, does not lose its lining. Its wall is much less muscular than the wall of the upper part of the uterus and contains more fibrous tissue. Its secretion provides a suitable medium for the spermatozoa to swim through but in some women the secretion may be too thick for the spermatozoa and cause sterility.

THE VAGINA. The *vagina* is a tubular structure with its anterior and posterior walls normally in apposition. It is about 8 cm long and extends from the cervix of the uterus to the exterior. It lies behind the bladder and in front of the rectum, and the urethra runs downwards in its anterior wall (Fig.97b). It passes downwards and forwards at an angle of 60° to the horizontal.

Its lower end is surrounded by two folds of skin on each side (Fig.108). The smaller inner folds are called the *labia minora.* They do not contain any fat. The larger outer folds are called the *labia majora.* They contain fat and have hairs on their outer surface. The labia majora are equivalent to the scrotum in the male.

In front of the labia minora is the *clitoris* which is the homologue of the penis in the male. In the virgin there is a thin fold of mucous membrane within and just above the labia minora. This is called the *hymen.* The opening in the hymen is of variable size and shape.

The vagina is lined by stratified squamous epithelium, and its wall consists of smooth muscle.

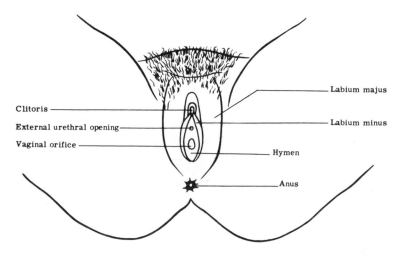

Fig.108. External female genitalia.

THE MAMMARY GLAND (BREAST). This is a glandular structure which develops from the skin, and its secretory tissue is said to consist of specialized sweat glands. Before puberty the glands in the female, as in the male, are rudimentary and consist mainly of a small central projection, the *nipple*, and surrounding pigmented skin, the *areola*. There is very little underlying breast tissue. At puberty the female breast enlarges. Early on this appears as a protrusion of the areola due to the development of the underlying breast tissue. The breast enlarges but the nipple does not protrude until the later stages of its development. Puberty takes place at a variable time but the major part of the development of the gland takes about two years. Growth of the gland can continue until about the age of twenty. There is an extension of the breast tissue upwards and laterally towards the armpit.

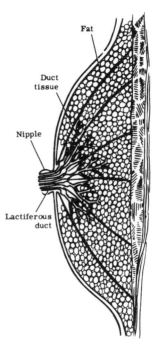

Fig.109. Longitudinal section of female breast.

In a woman who has never been pregnant most of the breast consists of connective tissue containing a variable amount of fat and its size depends largely on the amount of fat in the breast (Fig.109). The glandular tissue is mainly in the form of ducts which open on to the nipple. There is also a considerable amount of smooth muscle in the nipple and areola. The functions of this smooth muscle are to make the nipple protrude for feeding the baby and possibly to help the expression of the milk.

If a woman becomes pregnant the breasts enlarge. In the early months this is due to an increase in the duct tissue and in the later months due to an increase in the secretory tissue, the alveoli. The nipple and areola also become darker in pregnancy. In the later months of pregnancy and for about four days after the birth of the baby some secretion called *colostrum* is produced. This is not milk, which is secreted about the fifth day. Milk secretion continues for a variable time and in variable quantities. The stimulus of suckling and the regular emptying of the breasts are the most important factors in maintaining lactation. After the end of lactation the breast becomes reduced in size and most of the secretory tissue disappears, but there is always much more glandular tissue than in the breast of a woman who has never been pregnant.

After the menopause the breasts usually become smaller. This is associated with degeneration of the glandular tissue. In some women the breasts become enlarged due to the deposition of fat.

The menstrual cycle may be associated with changes in the breasts. In the second half of the cycle the breasts may become enlarged. This may be due to proliferation of the duct tissue but is usually due to accumulation of fluid (*fluid retention*) in the breast.

11

The nervous system

General

All cells possess the properties of excitability and conductivity. The nervous tissue of a complex animal is the tissue in which these properties are specially developed (see pp. 23 and 56). The neurone, consisting of a cell body and processes, and the synapse have already been described. The nervous system of the body is conveniently divided into *peripheral* and *central*. The peripheral nervous system consists mainly of bundles of nerve fibres which are in the form of nerves. These fibres are both sensory and motor, that is they convey impulses from the surface and deeper parts of the body towards the central nervous·system and impulses from the central nervous system to the muscles and glands of the body. The central nervous system is the means whereby impulses due to changes in the external and internal environment of the body are sorted out and correlated so that the appropriate response (or lack of response— *inhibition*) ensues. The central nervous system consists of the *brain* and *spinal cord* (Fig.110a).

THE NERVE IMPULSE. The stimulus which results in an impulse being propagated along a nerve fibre may be mechanical, electrical or chemical but specialization in the receptors occur so that each receptor responds more readily to a special type of stimulus, for example the receptor of the eyeball responds to a light stimulus. Receptors, however, respond to another type of stimulus if it is strong enough. A blow on the eye can give the impression of flashes of light. In general the stimulus has to be of sufficient strength, applied sufficiently quickly and last for a sufficient time before the neurone responds, that is, the stimulus must be of *threshold value*. If, however, *subthreshold stimuli* are applied frequently enough these

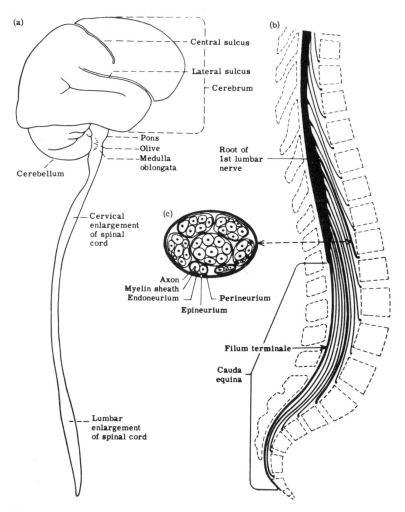

Fig.110. (a) Central nervous system (filum terminale has been omitted), (b) Formation of cauda equina (1st lumbar and 1st sacral nerve roots have been deliberately emphasized), (c) Transverse section of a nerve to show connective tissues sheaths.

stimuli can summate and produce a response. A stimulus in excess of threshold does not produce any greater effect. In other words the neurone obeys the *all or none law*.

The velocity of an impulse varies from 1 metre per second to 120 metres per second. This is related to the size and myelination of the fibres. Small, non-myelinated fibres conduct slowly and large, myelinated fibres, such as large motor fibres, conduct quickly.

WHITE AND GREY MATTER. Throughout the nervous system nerve fibres form bundles and cell bodies form aggregations. These are called respectively *white* and *grey matter*. It is important to note that in various parts of the nervous system grey and white matter have different names and even special names. A collection of cell bodies is sometimes called a *ganglion* or a *nucleus*, for example, the *posterior root ganglion* on the posterior root of a spinal nerve or the *caudate nucleus* of the forebrain. The *thalamus* is a large mass of cell bodies. It is obvious that grey matter contains the beginnings and ends of nerve fibres as well as cell bodies.

White matter looks white because of the presence of myelin (see p. 26) and also has different names, for example, a peripheral nerve consists of nerve fibres, as does a *tract* or *column* or *fasciculus* in the spinal cord or a *peduncle* or *commissure* or *lemniscus* in the brain.

DEGENERATION AND REGENERATION. It has already been stated that the structural unit of the nervous system is the neurone. If the cell body of a neurone is destroyed all its processes disappear and the neurone cannot be replaced. Neurones have no centriole and cannot divide so that neurones which are destroyed are not replaced.

A typical peripheral nerve consists of sensory and motor fibres of varying size and thickness of myelin sheath. Their cell bodies are in the posterior root ganglion or anterior horn of the spinal cord respectively (Fig.20). The individual nerve fibres are enclosed in a connective tissue sheath (the *endoneurium*) and the fibres are in bundles, each bundle being surrounded by the *perineurium*. The whole nerve is surrounded by the *epineurium* (Fig.110c). Both the perineurium and epineurium consist of connective tissue.

After a nerve is cut, changes take place in the peripheral stump, that is, in the axons and their sheaths which have been separated from,the rest of the nerve. The axons, although they can conduct an impulse for about twenty-four hours, break up and disappear in about four days. The myelin sheaths break up in two or three days and most of them have disappeared in about three weeks. Traces of myelin, however, can be detected for months afterwards. The Schwann cells multiply and fill the endoneurial tubes. These cells and other phagocytes are responsible for the removal of the axon and the myelin. The changes just described are known as *Wallerian degeneration*. The end result is that the peripheral stump consists of endoneurial tubes full of Schwann cells. The central stump is unaffected except for the area next to the cut. This area shows similar changes to those seen in the peripheral stump.

The cutting of a nerve results in the paralysis of the muscles and loss of sensation (inability to appreciate pain, changes of temperature and touch) in the skin or mucous membrane supplied by the nerve. The loss of sensation is not as extensive as would be expected because there is an overlap in the areas of skin supplied by two adjacent nerves. The paralysis of the muscles is called flaccid, because the muscles are said to be hypotonic, tendon reflexes involving the muscles are lost and the paralysed muscles waste.

The cell bodies of the nerve fibres which are cut also undergo certain changes called *chromatolysis*. In this the nucleus become eccentric and the Nissl's granules disappear. These changes reach their maximum in about five days and recovery takes place in about twenty-one days—the nucleus becomes central and the Nissl's granules re-appear.

Regeneration of a cut peripheral begins within hours or days. The axons grow out from the central stump across the cut towards the peripheral stump. If they enter the endoneurial tubes they grow towards the muscles and skin supplied by the nerve. Growth is at the rate of about 1 mm per day. After reaching the muscle or skin the fibres become myelinated and function returns. The motor units recover in succession and gradually the muscle recovers its power. Early on, the returning sensation in the skin is described as dull and ill-defined and subsequently as sharp and accurately located. Obviously some sensory fibres end in muscles (motor end plates) and some motor fibres end in sensory end organs or freely in the skin. These will not function. This explains to some extent why full recovery is unlikely. Other factors influencing the extent of recovery are

    a. the size of the gap, because axons cannot grow across an interval greater than about 2 cm,

    b. the state of the tissue in the gap because if this is dense the axons cannot grow through it,

    c. the extent to which the muscles have degenerated when they are eventually re-innervated.

In the central nervous system, nerve fibres separated from their cell body undergo similar changes to those seen in a peripheral nerve—the axon disappears, the myelin sheath disappears and cells, similar to the Schwann cells, increase in number. There are no endoneurial tubes, however, in the brain and spinal cord. The result is that although the axons attached to the nerve cells grow they become enmeshed in a network of processes of the cells forming the neuroglia and never reach their former terminations. No functional recovery takes place.

The spinal cord

The spinal cord is a nearly cylindrical structure about 45 cm long and about 1 cm wide. It lies in the vertebral canal and extends from the foramen magnum to the level of the intervertebral disc between the first and second lumbar vertebrae. Above, it is continuous through the foramen magnum with the *medulla oblongata* of the hindbrain, and below, it tapers rapidly to form the *filum terminale*, consisting of fibrous tissue and fusing with the back of the coccyx (Fig.110c). The cord is enlarged in the lower cervical and lumbar regions. These enlargements correspond with the upper and lower limb nerve plexuses respectively.

Arising from the anterolateral and posterolateral aspects of the cord on each side are two series of *rootlets, anterior* and *posterior* (Fig.20). These unite and form *anterior* and *posterior roots*. The *spinal nerves*, of which there are thirty-one pairs, are formed by the union of an anterior and posterior root. The anterior root is motor, the posterior root is sensory and the spinal nerve is mixed. On the posterior root is a swelling called the *posterior root ganglion* which contains the cell bodies of the sensory fibres of the spinal nerve. The peripheral process of a sensory neurone is in a spinal nerve and the central process is in a posterior root and goes to the spinal cord in one of the rootlets (Fig.111). The anterior and posterior roots become more and more oblique as one passes down the cord until the filum terminale is surrounded by nerve roots going towards their appropriate foramina. This results in the formation of the *cauda equina* (Fig.110b) in the lowest part on the vertebral canal. Early in fetal life the spinal cord is the same length as the vertebral canal and there is no cauda equina. Subsequently the vertebral column grows in length much more rapidly than the spinal cord and as a result the lower nerve roots become oblique instead of horizontal.

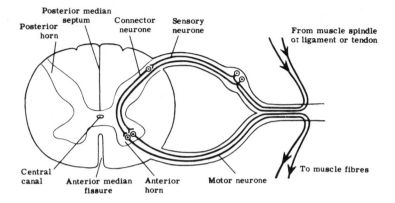

Fig.111. Section of spinal cord and formation of a spinal nerve in which are shown the basic neurones of either a monosynaptic or disynaptic reflex arc.

The spinal cord has an *anterior median fissure* which is about 3 mm deep. In cross-section the spinal cord is seen to consist of an internal fluted column of grey matter surrounded by white matter (Fig.111). The grey matter is H-shaped and the transverse part of the H contains the small *central canal*. The lateral limbs of the H are symmetrical and crescentic and are divided by the transverse limb into *anterior* and *posterior horns*. In the thoracic region there is an additional *lateral horn* of grey matter between the anterior and posterior horns. The posterior horn is narrow and usually reaches nearly to the surface of the cord. The anterior horn is bulbous and does not reach the surface. There is a *posterior median septum* in the midline posteriorly extending from the surface of the cord to the grey matter. This together with the anterior median fissure divides the cord into two halves. Each half is further divided into three by the entry or emergence of the rootlets. The *posterior column* of white matter is between the septum posteriorly and the posterior horn, the *lateral column* is between the two horns and the *anterior column* is between the anterior horn and the anterior fissure (Fig.115).

THE REFLEX ARC. The functional unit of the nervous system, the *reflex arc*, can now be described. Basically, it consists of:

   a. a *sensory neurone* with a peripheral process ending in a receptor organ or freely in the skin, a cell body in the posterior root ganglion and synapsing with

b. a *motor neurone* with a cell body in the anterior horn and an axon which leaves the spinal cord in the anterior root and ends in an effector organ such as a muscle (Fig.111).

If the receptor is stimulated the effector will respond. This, however, is much too simplified. Frequently there is a neurone between the sensory and motor neurones, the *intercalated (connector) neurone.* The situation in which one sensory neurone makes contact with one intercalated neurone which makes contact with one motor neurone does not really exist. One sensory neurone usually makes contact with many intercalated neurones which make contact with many motor neurones, or the impulse along one sensory neurone reaches more than one motor neurone by different paths. The branching of the axons in the central nervous system results in enormously complex interconnexions between neurones.

What is known as a *tendon reflex* is an example of the functioning of the reflex arc. If a tendon such as the patellar ligament (really the tendon of the quadriceps femoris muscle, the large muscle on the front of the thigh) is stretched quickly enough, the stretch receptors in the ligament are stimulated and impulses travel along the afferent (sensory) neurones to the spinal cord. These neurones synapse with anterior horn cells whose axons pass to the quadriceps femoris muscle which contracts so that the leg is raised at the knee. It should be pointed out that even this simple reflex involves more than one segment of the spinal cord. The sudden stretching of almost any tendon will produce reflex contraction of its muscle. Similarly, stroking the skin of the abdominal wall frequently produces the reflex contraction of the abdominal muscles. This is known as a *skin reflex.* Other examples of reflex activities are seen in temperature control, changes in heart rate and blood pressure, and blinking.

THE GENERAL ARRANGEMENT OF SENSORY AND MOTOR PATH- WAYS IN THE CENTRAL NERVOUS SYSTEM. Although many functions of the body are carried out at a reflex level by means of neuronal connexions not unlike those described in the reflex arc (a sensory neurone linking with a motor neurone with or without intervening connector neurones so that sensory information produces motor activity), many sensory impulses reach the highest levels of the central nervous system and many motor activities are initiated from the same part of the nervous system. The spinal cord can be regarded in some ways as a simple part of the central nervous system and the brain as a more complex part. The brain itself has different degrees of complexity and is conveniently divided into the *hindbrain, midbrain* and *forebrain*, the latter being the most complex and in man the largest part of the brain (Fig.119). The cerebral hemispheres constitute the major part of the forebrain. This increased size of the forebrain is associated with a great extension of the functions of the nervous system which include all the intellectual activities of which man is capable, and also in an increased control of the lower parts of the nervous system by the forebrain.

Although many of the co-ordinated activities of the body do not involve the cerebral hemispheres it is convenient to think of sensory pathways ending in the surface grey matter of the hemispheres (the *cerebral cortex*), and the motor pathways beginning in the cerebral cortex. Sensory pathways usually have three neurones (Fig.112). The first neurone is either pseudo-unipolar or bipolar with its cell body outside the central nervous system (in a ganglion of either a cranial nerve or a posterior nerve root) and a peripheral process going to a receptor end organ or ending freely in the skin or a mucous membrane. Its central process enters the brain or spinal cord and ends by synapsing with the cell body of the second neurone at the same or a different level. The axon of the second neurone usually crosses the midline and goes to the thalamus, a mass of grey matter in that part of the forebrain immediately above the midbrain. The third neurone has its cell body in the thalamus and its axon passes through the central part of the cerebral hemisphere to the cerebral cortex.

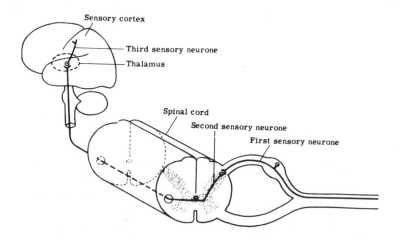

Fig.112. General arrangement of sensory pathway in central nervous system.

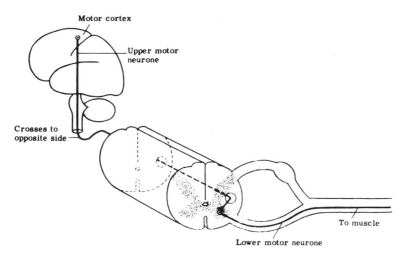

Fig.113. General arrangement of motor pathway in central nervous system.

The motor pathway from the cerebral cortex has two neurones, an *upper motor neurone* and a *lower motor neurone*. The fibres of the former form the *pyramidal tract* or *fibres*. An upper motor neurone has a cell body in the motor cerebral cortex and a descending axon which crosses the midline usually in the hindbrain (sometimes in the spinal cord) and ends in relation to the anterior horn cell (Fig.113). The lower motor neurone has its cell body in the anterior horn of grey matter of the spinal cord and an axon which leaves the spinal cord in the anterior root, runs in a spinal nerve and ends in a muscle. The pyramidal fibres are not the only motor neurones which act on the anterior horn cell. *Extrapyramidal fibres* which come from the midbrain and hindbrain, and are influenced by masses of grey matter near the thalamus in the forebrain, also synapse with the anterior horn cells. The lower motor neurone, however, is the only route whereby an impulse from the central nervous system can reach a muscle, and is referred to as the *final common pathway* (Sherrington).

Two further points may be made. The motor cranial nerves are also acted on by pyramidal and extrapyramidal fibres. Secondly, it can be seen that on the whole the sensory pathways of one side of the body end on the opposite side of the forebrain, and the motor cortex on one side of the forebrain acts on the lower motor neurones of the other side of the spinal cord.

RECEPTORS. It has already been pointed out that the central nervous system receives information about the environment and the body itself by means of sensory neurones. The peripheral process of these neurones is usually related to a special structure which is called a *receptor end organ*. The peripheral processes of some sensory neurones have free nerve endings and it is assumed that these are stimulated by direct contact. Receptor end organs are sensitive to a specific stimulus, for example the rods and cones of the retina of the eyeball respond to light stimuli, but all will react if a stimulus is strong enough.

Table 5

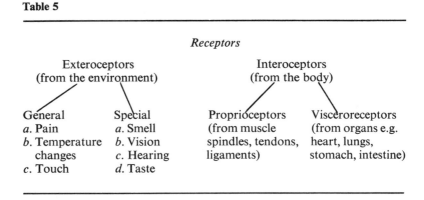

The receptors are subdivided in the following way (Table 5). *Exteroceptors* react to environmental stimuli and are subdivided into *general* (pain, temperature and touch) and *special* (vision, hearing, taste and smell). *Interoceptors* react to stimuli from within the body and are subdivided into *proprioceptors* (from muscles, tendons and ligaments, and the semicircular canals, etc. of the internal ear) and *visceroreceptors* (from organs such as the heart, lungs, stomach, intestine, bladder). There are no specific end organs associated with the sensation of pain. An impulse which is interpreted as painful begins in fine, free nerve endings in the skin or mucous membrane. Impulses interpreted as warm or cold may have specific structures in the dermis of the skin and these respond to the specific stimulus of warmth or cold. Touch is associated with *Meissner's corpuscles* in the dermal papillae, and also the movement of the hairs, and pressure with *Pacinian corpuscles* in the deeper parts of the dermis (Fig.114).

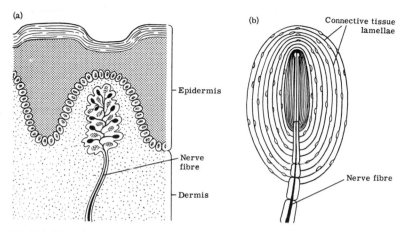

Fig.114. (a) Meissner's corpuscle (touch receptor end organ), (b) Pacinian corpuscle (pressure receptor end organ).

The special senses also have specific end organs which respond to different types of stimuli. The proprioceptive end organs are the muscle spindles, (see Fig.15) *neurotendinous end organs* (in tendons near their attachment to muscles) and Pacinian corpuscles (in ligaments).

THE TRACTS OF THE SPINAL CORD. Apart from a few differences in the cervical region, sections of the spinal cord at all levels look similar (Fig.115). There is an internal H-shaped arrangement of the grey matter surrounded by white matter. The fibres in the white matter are, however, arranged in definite bundles or tracts and these tracts are related to whether they are descending (motor) or ascending (sensory). In addition the fibres associated with one type of sensation,

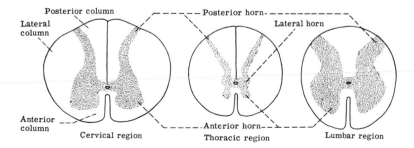

Fig.115. Appearance of transverse sections of spinal cord at different levels.

for example, pain, form one tract, and motor fibres associated with one particular function remain together. As a result it is possible to indicate the position of various tracts with some precision.

Fibres whose impulses are interpreted in terms of pain and changes of temperature enter the spinal cord via the posterior roots and synapse with cells in the posterior horn of grey matter (Fig. 112). Their cell bodies are in the posterior root ganglia and their peripheral processes end in the skin, etc. The second neurones have cell bodies in the posterior horn and fibres which cross the midline and run upwards in the lateral column of white matter through the hindbrain and midbrain to the thalamus. Because of this it is called the *lateral spinothalamic tract*. The third neurone corresponds with the description in the general plan of sensory neurones. It has a cell body in the thalamus and a fibre which goes to the cerebral cortex.

The pathway mediating impulses interpreted as touch has a similar arrangement. The first neurones are pseudo-unipolar with their cell bodies in the posterior root ganglia. Their peripheral processes end in relation to hair follicles or Meissner's corpuscles in the skin and their central processes end in the posterior horn where they synapse with the second neurone. The second neurones have cell bodies in the posterior horn and fibres which cross the midline and run upwards in the anterior column of white matter to the thalamus, hence their name, *anterior spinothalamic tract*. The third neurone is similar in its arrangement to other third neurones.

Pathways for proprioception fall into two groups. In the first, often referred to as *conscious* proprioception, there are three neurones. The first neurones are pseudo-unipolar with cell bodies in the posterior root ganglia, peripheral processes which end in relation to muscle spindles or proprioceptive end organs in tendons or ligaments, and central processes which enter the spinal cord in posterior nerve roots and run upwards on the same side in the posterior columns as the *fasciculus gracilis* (medial) and the *fasciculus cuneatus* (lateral). These fibres end in nuclear masses in the hindbrain where they synapse with the cell bodies of the second neurones. The axons of the second neurones cross the midline and run upwards through the hindbrain and midbrain to the thalamus where they synapse with the third neurone whose arrangement has been described. It should be noted that although the spinal cord contains the *uncrossed*, central processes of the first neurones, the general arrangement of the three neurones is the same as is found in the other sensory pathways (Fig. 116).

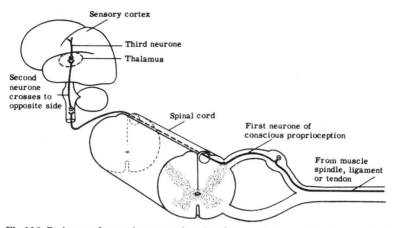

Fig.116. Pathway of conscious propriception from periphery of body to cerebral cortex.

The second proprioceptive pathway, often referred to as *unconscious* because it ends in the cerebellum and not the cerebral cortex, is quite different. It has two neurones and the second is as a rule uncrossed (Fig.117). The first neurones have cell bodies in posterior root ganglia, peripheral processes ending in relation to end organs in muscles, tendons or ligaments, and central processes which end in the posterior horn. The second neurones have cell bodies in the posterior horn and axons which run upwards in the lateral column

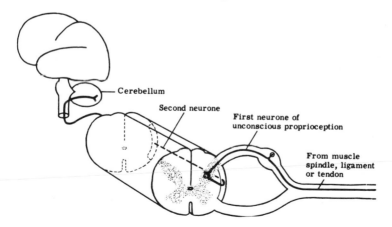

Fig.117. Pathway of unconscious propriception from periphery of body to cerebellum.

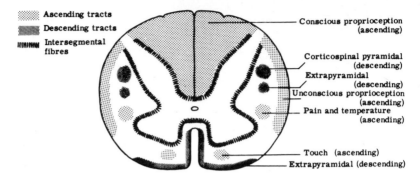

Fig.118. Transverse section of spinal cord to show position of main tracts of fibres.

of white matter of the same side to the cortex (grey matter) of the cerebellum. They are therefore called the *spinocerebellar tracts*. It should be added that touch is often divided into light touch and crude touch. The pathway of light touch is the same as that of conscious proprioception and that of crude touch is the one which includes the anterior spinothalamic tract.

A knowledge of the position of the tracts in the spinal cord (Fig.118) helps one to understand the signs and symptoms of certain diseases and injuries of the spinal cord. For example, degeneration of the posterior columns, seen in *tabes dorsalis*, a late complication of syphilis, results in the patient not having adequate information about the position of his lower limbs when walking because this information comes from the muscles, tendons and ligaments of his lower limbs and passes to the brain in the posterior columns.

Before considering the position of the descending tracts in the spinal cord the term *modalities of sensation* should be explained. It refers to the different types of sensation capable of being appreciated by a human being, such as pain and touch. The appreciation of these differences is a function of the brain. The impulses which travel along neuronal pathways are all identical. Only when the impulse reaches the brain are the different types of sensation recognized.

A term commonly used as synonymous with proprioception is *kinaesthesia*, which means *a feeling of movement*. By means of information from the muscles, tendons and ligaments one regulates the type, force and speed of movement required to carry out any activities including everyday activities such as walking. In addition a knowledge of the shape, size, weight and texture of an object is obtained from proprioceptive pathways because this information depends on the position of the fingers and the segments of the limb, and the force with which muscles are contracted. This is often referred to as *stereognosis*.

The pyramidal and extrapyramidal tracts are also arranged in definite bundles in particular parts of the white columns. The pyramidal tracts are called *corticospinal* for obvious reasons. The majority of the fibres are crossed and are found in the lateral column (*lateral corticospinal tract*). The minority are uncrossed and are found in the anterior column (*anterior corticospinal tract*). They cross to the opposite side before synapsing with the anterior horn cells. The extrapyramidal, of which there are several, are found in the lateral column (*rubrospinal tract* from the midbrain) and anterior column.

There are other tracts in the spinal cord in addition to the tracts described and also fibres which come from cells in the grey matter, enter the white matter and run up or down for a variable distance before re-entering the grey matter. The latter are called *intersegmental fibres* and are situated near the grey matter.

The hindbrain

This consists of three parts

- *a.* the *medulla oblongata* which is about 2·5 cm long and is the upward continuation of the spinal cord through the foramen magnum,
- *b.* the *pons* which is above the medulla and is about 3·5 cm wide and 3·0 cm long,
- *c.* the *cerebellum* which is a much larger structure (10 cm wide, 5 cm long and 4 cm thick) and lies behind the medulla and pons (Fig.119).

Above the hindbrain is the midbrain. Externally on the anterior (inferior) aspect of the medulla is the *pyramid* on either side of the midline, and lateral to the upper part of the pyramid is the *olive*, an oval structure about 1 cm long. Lateral to the olive is the *inferior cerebellar peduncle* (Fig.120). Between the pyramid and olive emerge the rootlets of the *hypoglossal* (12th cranial) *nerve*, and between the olive and the inferior peduncle are the cranial part of the *accessory* (11th cranial), the *vagus* (10th cranial) and the *glossopharyngeal* (9th cranial) *nerves* (Fig.119).

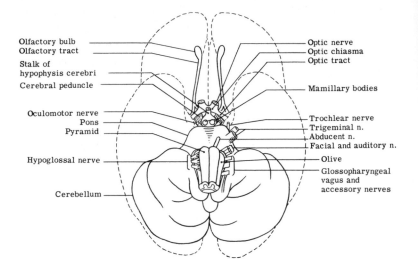

Olfactory bulb
Olfactory tract
Stalk of
hypophysis cerebri
Cerebral peduncle

Optic nerve
Optic chiasma
Optic tract

Mamillary bodies

Oculomotor nerve
Pons
Pyramid

Trochlear nerve
Trigeminal n.
Abducent n.
Facial and auditory n.

Hypoglossal nerve

Olive

Cerebellum

Glossopharyngeal
vagus and
accessory nerves

Fig.119. Inferior surface of brain.

On the posterior aspect of the medulla (Fig.120) the posterior columns of the spinal cord can be seen ending on either side of the midline in swellings in which lies the grey matter containing the synapses between the first and second neurones of the conscious proprioceptive pathway. This region is also the inferior angle of the diamond-shaped floor of the *fourth ventricle*, the dilated part of the central canal of the hindbrain (Fig.121). The roof of the ventricle is tent-like and is covered by the middle part (*vermis*) of the cerebellum. The lower sides of the diamond are formed by the diverging inferior cerebellar peduncles and the upper sides by the converging *superior cerebellar peduncles*. At the lateral angles of the diamond on each side are the three cerebellar peduncles. In the floor of the fourth ventricle are various symmetrical elevations which are due to underlying small masses of grey matter associated with the cranial nerves.

The pons superficially consists of transverse fibres which laterally form the *middle cerebellar peduncles*. More deeply, the pons is the upward continuation of the medulla. At the lower border of the pons laterally are the *auditory* (8th cranial) and *facial* (7th cranial) *nerves* and nearer the midline, the *abducent* (6th cranial) *nerve*. Where the pons becomes the middle peduncle the *trigeminal* (5th cranial) *nerve* is seen (Fig.119).

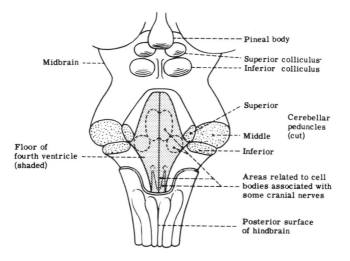

Fig.120. Posterior surface of hindbrain and midbrain after removal of cerebellum.

The cerebellum consists of two lateral hemispheres and a central part, the vermis. All three parts consist of an outer, much folded cortex and inner white matter, although there are several nuclei (grey matter) in the white matter in or near the midline. All fibres go to or come from the cerebellum in the *peduncles*. Broadly speaking the *inferior* and *middle peduncles* contain ingoing fibres, that is, fibres from cell bodies outside the cerebellum, and the *superior* contain outgoing fibres, that is, fibres from cell bodies inside the cerebellum. The inferior peduncles contain fibres from end organs in muscles, tendons and ligaments (proprioceptive) via the spinal cord, fibres from the vestibular nuclei (proprioceptive) and fibres from the grey matter in the olive. The middle peduncles contain fibres from nuclei in the pons. These nuclei receive the terminations of fibres whose cell bodies are in the cerebral cortex of the forebrain. The fibres in the superior peduncles leave the cerebellum and go to the midbrain and thalamus. The midbrain is connected with the anterior horn cells of the spinal cord and the thalamus is connected with the motor areas of the cerebral cortex. The motor areas are connected with the anterior horn cells. In this way the cerebellum receives information from structures associated with movement (muscles and joints) and co-ordinates movements through the nerve supply of muscles.

The internal structure of the medulla and pons shows a similarity to the spinal cord in its lowest part but higher up the changes are marked. These changes are due to the presence of the nuclei (cell bodies of neurones) associated with cranial nerves and structures such as the olive, and the crossing of fibres from one side to the other, for example, the crossing of the pyramidal fibres (descending). In general it may be said that there are

a. ascending and descending tracts, for example the spino-thalamic tracts (ascending) and the pyramidal tract (descending),

b. cranial nerve nuclei associated with the 12th to the 5th cranial nerves,

c. small masses of grey matter associated with reflex, vital activities such as the heart rate, respiration and blood pressure (cardiac, respiratory and vasomotor centres).

The cerebellum and the superficial part of the pons are associated with the co-ordination of movements. Sensory impulses from muscles, tendons and ligaments, and information about the position of the head in space are fed into the cerebellum mainly through the inferior peduncles. The cerebral cortex, through the nuclei of the pons and their fibres in the middle peduncles, monitors the activity of the cerebellum which acts on various parts of the nervous system through the outgoing fibres in the superior peduncles. Disturbance of the cerebellum is evidenced by disturbance of co-ordinated motor functions. For example, walking and speech are affected and there is an inability to hold the eyes steadily in a lateral direction. The cerebellum does not initiate movement and its removal does not produce paralysis of muscles or any sensory disturbance.

The midbrain

This is the smallest of the three subsections of the brain and lies above the pons. It is about 1·5 cm long, 2·5 cm wide and 2 cm thick. Anteriorly it consists of the diverging *cerebral peduncles* and posteriorly there are four small elevations arranged in two pairs, the *superior* and *inferior colliculi* (Fig.120). Between the cerebral peduncles is the *interpeduncular fossa* which forms the floor of the *third ventricle* of the forebrain (Fig.121). In this fossa from behind forwards are the paired *corpora mamillaria*, a central area to which is attached the *stalk of the hypophysis cerebri* (*pituitary gland*) and the *optic chiasma* in the midline. The transverse section of the midbrain

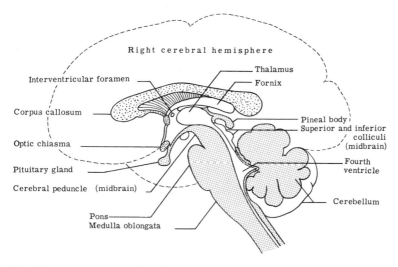

Fig.121. Sagittal section of brain.

shows the *cerebral aqueduct* nearer the posterior than the anterior aspect. This is a narrow channel connecting the fourth ventricle with the third ventricle (Fig.121).

In the midbrain are ascending and descending tracts of fibres and the nuclei of the *trochlear* (4th cranial) and *oculomotor* (3rd cranial) *nerves*. The trochlear nerves emerge lateral to the cerebral peduncles and the oculomotor nerves medial to the peduncles. There are also nuclear masses associated with the extrapyramidal system, for example the *red nucleus*. The inferior colliculi are associated with auditory reflexes and the superior with visual reflexes.

The forebrain

As has already been stated, the cerebral hemispheres form the major part of the forebrain. These develop as lateral extensions of the primitive neural tube. That part of the forebrain which represents the original head end of the neural tube and lies immediately above the midbrain is called the *diencephalon*. The lateral extensions form the cerebral hemisphere by growing forwards (*frontal lobes*), backwards (*occipital lobes*) and downwards and then forwards (*temporal lobes*). The hemispheres thus overgrow the rest of the brain which can be easily seen only from below (Fig.122).

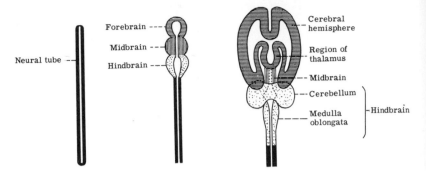

Fig.122. Development of central nervous system.

THE THALAMUS. The diencephalon consists mainly of the two thalami which lie one on either side of the midline (Fig. 121). Between them is the *third ventricle*, an enlargement of the central canal of the neural tube. In front of the thalamus on each side is the *interventricular foramen* which is an opening between the third ventricle and the cavity of the cerebral hemisphere, the *lateral ventricle*. The thalamus itself is a mass of grey matter, shaped like an egg, with the smaller end lying anteriorly. It is about 3 cm long, 1 cm wide and 1 cm high. Posteriorly it projects over the midbrain. Lateral to the thalamus is the *posterior limb* of the *internal capsule* which separates it from the *lentiform nucleus* (see below). Above it are the *fornix* and the posterior part of the *corpus callosum* (Fig. 121). Below it, near the midline, is the *hypothalamus*, an area which corresponds with the floor of the third ventricle between the thalami. More laterally the upper part of the midbrain lies below. At the back of the thalamus are areas associated with visual and auditory pathways.

The thalamus itself is divided into a large number of nuclear masses with different functions and connexions (Fig. 123). It will be recalled that the axons of the second neurones of the sensory pathways end in the thalamus. From the thalamus (lateral part) the axons of the third neurones go to various parts of the cerebral cortex (general sensation to the postcentral gyrus, vision to the occipital lobe and hearing to the temporal lobe). Secondly the medial part of the thalamus is connected with the hypothalamus, and through this influences autonomic function. Thirdly the cerebellum is connected with the thalamus which in turn is connected with the motor cortex. In this way motor functions are influenced by the

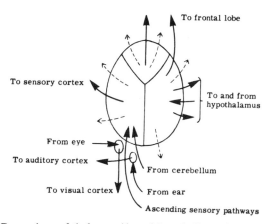

Fig.123. Connexions of thalamus (dotted lines indicate that whole thalamus has connexions with whole cerebral cortex).

thalamus. The thalamus is also connected with the frontal lobe, and these connexions are associated with the personality of an individual especially in relation to aggression and mental tension. The thalamus also has diffuse connexions with the whole cerebral cortex and this is thought to be associated with the functions of alertness and consciousness.

The parts of the forebrain round the thalamus are named according to their relation to the thalamus. The *hypothalamus* lies below the medial part of the thalamus and consists of several groups of cell bodies and tracts of fibres near the optic chiasma, near the attachment of the stalk of the hypophysis cerebri and near the mamillary bodies. The hypothalamus regulates certain vital functions, such as fat and carbohydrate metabolism, temperature regulation, hunger and thirst, and sleep. It also influences the activities of the autonomic nervous system. Connexions between parts of the cerebral cortex and the hypothalamus are also related to emotional responses to certain stimuli. The cortex apparently damps down what would be an excessive emotional reaction to some stimuli.

THE CEREBRAL HEMISPHERES. Each hemisphere has three surfaces—a flat, vertical, medial surface, a convex, superolateral surface, and a somewhat irregular, inferior surface. The hemisphere is divided into lobes called after the main bone to which each is related. These lobes are demarcated by some of the grooves on the

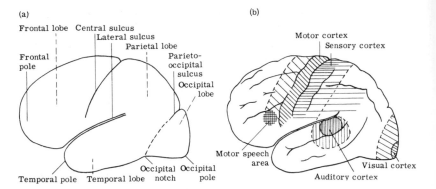

Fig.124. (a) Subdivisions of cerebral hemisphere, (b) Functional areas on lateral surface of cerebral hemisphere (wide hatching indicates associative areas).

surface of the hemisphere (Fig.124). The grooves are called *sulci* and the areas between grooves are called *gyri*. On the superolateral surface a horizontal fissure (*lateral sulcus*) is seen extending backwards below the middle of the surface for about 8 cm. Anteriorly it extends medially through the whole thickness of the hemisphere. The *central sulcus* runs downwards and forwards on this surface beginning at the superior border and ending just above the lateral sulcus. The *frontal lobe* lies in front of the central sulcus and above the lateral sulcus. Its anterior rounded part is called the *frontal pole*. There is on the superior border, about 5 cm behind the central sulcus, the upper part of the *parieto-occipital sulcus* which is on the medial surface of the hemisphere. The *occipital notch* is on its lower border about 5 cm in front of the rounded posterior end of the hemisphere (*occipital pole*). The *occipital lobe* lies behind a line drawn between the parieto-occipital sulcus and the occipital notch. The *temporal lobe* lies in front of this line and below the lateral sulcus, and the *parietal lobe* is in front of this line, above the lateral sulcus and behind the central sulcus. The rounded anterior end of the temporal lobe is called the *temporal pole*.

The cerebral cortex. Most of the hemisphere consists of an outer folded *cortex* consisting of grey matter, and inner white matter consisting of nerve fibres. There are, however, masses of grey matter lateral to and in front of the thalamus and also in the temporal lobe. Apart from these the major part of the hemisphere consists of outer grey and inner white matter.

As has already been stated the cortex is very much folded and has even grown over and buried an area of cortex which can be seen if the edges of the lateral sulcus are pulled apart. This area is called the *insula*. The cerebral cortex is about 1 to 4 mm thick and consists of enormous numbers of neurones, many of which receive large numbers of fibres from sensory pathways. Many of the fibres from these neurones form motor pathways. In addition there are innumerable connexions between the neurones within the cortex. The result is that the cortex can be regarded as a receptor area with properties which make possible the correlation of sensory impressions as well as a means whereby voluntary movement can be controlled. Memory and abstract thought are also possible because of the enormous complexity of the neuronal arrangement. Many attempts have been made to classify the various areas of the cortex according to their structure, but it is difficult to correlate these structural differences with precise functional differences. The cortex is often divided into six layers of cells with two layers of fibres running horizontally between some of the layers of cells, but there is great variation in this arrangement.

Functional cortical areas. Some areas of the cortex are associated with fairly definite functions. About 1 cm in front of and parallel to the central sulcus is the *precentral sulcus* and between the two sulci is the *precentral gyrus* (Fig.124). This is called the *motor area* or *cortex*. This cortex is twice as thick as the *sensory area* or *cortex* which lies behind the central sulcus. In the fifth layer from the surface of the motor cortex are the *giant pyramidal cells of Betz* whose fibres form part of the pyramidal tract. (It is important to realize that this is not the only motor area of the cortex and the giant cells of Betz account for only a proportion of the fibres of the pyramidal tract.) The muscles of the body are represented upside-down on the motor cortex, that is the muscles of the head are at the lower end, and the muscles of the lower limb at the upper end. The area of cortex associated with any group of muscles is not related to the size of the muscles. For example, the muscles of the lips and the hand have a much bigger area than the muscles of the buttock. Furthermore movements are represented on the cortex, not individual muscles.

Running parallel to and about 1 cm behind the central sulcus is the *postcentral sulcus*. Between these two sulci is the *postcentral gyrus* which is the sensory area. This receives the terminal fibres of the third neurone of the sensory pathways of pain and temperature, touch and conscious proprioception. Again the body is represented upside-down on this gyrus and the area of cortex associated with an area of the body is not related to its size but to its importance in acquiring information about the environment. For example, the cortical sensory area for the hand is much greater than that for the foot.

In front of the lower end of the motor area in the left hemisphere in right-handed people is the *motor speech area (Broca's convolution)*. There are many areas of the cortex in addition to Broca's which are associated with various aspects of language (spoken, written, etc.). Many are near the lower ends of the motor and sensory areas. This is not surprising since these areas are related to the motor and sensory pathways connected with the muscles, skin and mucous membrane of the head and neck.

The *auditory area* is on the posterior end of the temporal lobe below the lateral sulcus. It receives the terminations of the third neurone of the auditory pathway mainly from the opposite ear. High notes are appreciated in the front part of the auditory cortex.

On the medial surface of the occipital lobe is the horizontal *calcarine sulcus* which extends backwards to the occipital pole (Fig. 125). The *visual cortex* lies on both sides of this sulcus. The visual cortex receives the terminations of the fibres constituting the visual pathway. The central area of the retina, called the *macula*, is projected on to a relatively large area of the cortex on the occipital pole. The right visual cortex receives impulses from the right halves of each retina. The result is that the two images of an object, one from each retina, are transmitted to the visual cortex of one side so that stereoscopic vision, the basis of perspective, is possible.

The area of the cortex associated with smell is situated on the medial side of the anterior part of the temporal lobe, and the area associated with taste may be in the lower part of the parietal or in the temporal lobe.

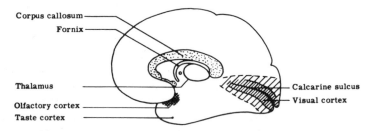

Fig. 125. Medial surface of cerebral hemisphere indicating areas associated with vision, smell and taste.

All the cortical areas related to specific functions are said to have *associative areas* adjacent to the areas described. For example the area in front of the motor area is also motor in function, the area below and in front of the auditory area is auditory in function and the area around the visual area is visual in function.

The white matter of the cerebral hemispheres. This consists of enormous numbers of nerve fibres and neuroglial cells. Some fibres connect gyri of the same hemisphere to each other, and play a very important role in the integration of the activities of different parts of the cortex but it is difficult to be more precise.

Fibres which cross the midline and connect the two hemispheres, usually corresponding areas of each hemisphere, are called *commissural*. The largest of these is the *corpus callosum* which roofs the lateral ventricles. In the midline it is about 8 cm long (Fig.121). Since the frontal lobes project forwards beyond the corpus callosum in front and the occipital lobes project backwards behind, the fibres passing from one frontal lobe to the other, or from one occipital lobe to the other, have to arch backwards or forwards respectively. The corpus callosum links the two hemispheres and enables information going to one hemisphere to be sent to the other as well. It also integrates the functions of the two hemispheres.

Some fibres go to or leave the grey matter of the cortex and connect it with the thalamus, midbrain, hindbrain and spinal cord. They therefore include ascending fibres going from the thalamus to the sensory, auditory and visual areas of the cortex, and descending fibres leaving the cortex and forming the pyramidal tract.

The basal ganglia. As the name suggests, this term refers to masses of grey matter inside the hemisphere in the base of the brain. They are sometimes called the *basal nuclei*. Occasionally the thalamus is included. The important nuclei are called the *caudate nucleus* and the *lentiform nucleus*. Together they are called the *corpus striatum* because the area between them looks striped due to a mixture of grey and white matter. The caudate nucleus, so-called because of its long tail, has a large head which is in front of the thalamus and a tail which arches backwards and then downwards and forwards into the temporal lobe. The lentiform nucleus is shaped like a biconvex lens and lies lateral to the thalamus (behind) and the caudate nucleus (in front) (Fig.126). Between the lentiform nucleus (laterally) and the caudate nucleus and thalamus medially is the *internal capsule* which consists of fibres passing between the cerebral cortex and the various parts of the brain and the spinal cord.

The corpus striatum together with the *red nucleus* of the midbrain has motor functions which are called *extrapyramidal* as distinct from the motor functions of the pyramidal fibres. In lower animals the corpus striatum is the highest motor centre but in man this function is taken over by the motor areas of the cortex. Disease of the corpus striatum is associated with tremor, weakness and rigidity of the muscles of the body. There is no real paralysis of the

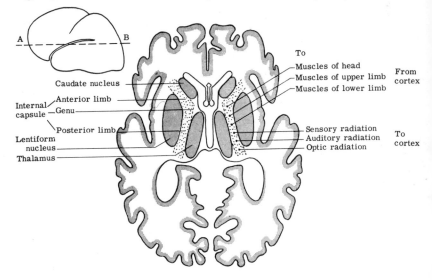

Fig.126. Horizontal section through cerebral hemispheres, as indicated, and appearance of surface of section.

muscles although there are disturbances of walking and posture due to the rigidity. This condition is seen in *paralysis agitans* (*Parkinson's disease*) and is often referred to as *Parkinsonism*. This condition usually occurs in older people (over 60 years). Disease of the other parts of the extrapyramidal structures is often associated with sudden involuntary movement such as twitching of the face (as seen in *chorea*) and grosser movements of the limbs (called *athetosis*).

The internal capsule. Reference has already been made to this structure which consists of a bunching together of fibres going to and coming from the cerebral cortex. The internal capsule is said to have two *limbs* set at an angle to each other (Fig.126). The angle is called the *genu*. The *anterior limb* lies between the caudate nucleus medially and the lentiform nucleus laterally. The *posterior limb* lies between the thalamus medially and lentiform nucleus laterally.

The fibres constituting the internal capsule are arranged in a certain order (Fig. 126). Running downwards through the genu and the adjacent area of the posterior limb are the pyramidal fibres from the motor cortex on their way to the cerebral peduncle of the midbrain. Within this area the fibres associated with face muscles are in the genu and next to these are the fibres associated with the upper limb muscles. Behind these are the fibres associated with the lower limb muscles. Next to these are ascending fibres (the third neurone of the sensory pathways) passing from the thalamus to the postcentral gyrus. This is called the *sensory radiation.* Behind these are the auditory fibres passing from the thalamus to the auditory area on the temporal lobe (*auditory radiation*) and behind these are the optic fibres passing from the thalamus backwards to the visual cortex on the medial side of the occipital lobe (*optic radiation*).

Cerebral haemorrhage frequently occurs in the region of the genu and adjacent part of the posterior limb of the internal capsule. The result is a *hemiplegia* (paralysis of the muscles of the opposite half of the body). The paralysis is a typical upper motor neurone paralysis with spasticity of the affected muscles, increased reflexes, absence of wasting of the muscles and a positive Babinski reflex (on stroking the sole of the foot the toes move upwards). The paralysis is usually much more extensive immediately after the haemorrhage since some of the fibres at the periphery of the haemorrhage are pressed on and will recover, whereas the fibres which are divided will never recover.

The membranes of the brain and spinal cord

There are three membranes round the brain and spinal cord called the *meninges* (Figs.127a,b). The outer is the *dura mater* which consists of tough fibrous tissue. It is mainly protective in function. The dura mater of the spinal cord continues downwards over the filum terminale and blends with it.

Within the skull there are said to be two layers of dura mater. The outer is really the periosteum of the skull bones so that the inner is the true dura mater and is continuous with the spinal dura through the foramen magnum. The cranial layers of the dura mater are in contact with one another except in those places where the venous sinuses run between the two layers. The inner layer forms partitions

(a)

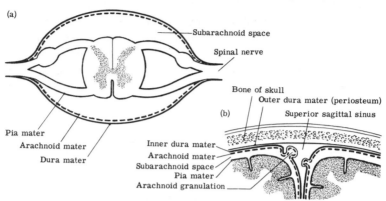

Fig.127. (a) Transverse section through spinal cord and meninges, (b) Coronal section through skull, cerebral hemispheres and meninges.

  a. between the cerebral hemispheres (*falx cerebri*)

  b. between the cerebral hemispheres above and the cerebellum below, in the posterior part of the cranial cavity (*tentorium cerebelli*) (Fig.128).

The superior sagittal sinus runs backwards in the upper border of the falx cerebri, and the inferior sagittal sinus in its lower border. The straight sinus runs in the junction between the falx cerebri and the tentorium cerebelli, and the transverse sinus, one on each side, runs in the outer border of the tentorium cerebelli.

  As the spinal and cranial nerves pass through the dura they carry

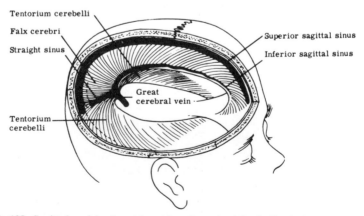

Fig.128. Sagittal and horizontal section through right half of skull and brain showing falx cerebri and tentorium cerebelli.

a sleeve of that membrane with them. This sleeve becomes continuous with the epineurium, the outer connective tissue covering of these nerves.

Lining the dura and adherent to it is the *arachnoid mater*, a very fine membrane which is associated with the re-absorption of the cerebrospinal fluid. Firmly attached to the brain and spinal cord is the third membrane (*pia mater*) in which the blood vessels of the brain and spinal cord run. Between the arachnoid and pia is the *subarachnoid space* containing the *cerebrospinal fluid*. It is important to recall that the spinal cord ends at the level of the first lumbar vertebra and to note that the subarachnoid space continues downwards to the level of the second sacral vertebra. By passing a hollow needle between the third and fourth lumbar vertebra it is possible to obtain a specimen of cerebrospinal fluid without the risk of damaging the spinal cord (*lumbar puncture*).

The cerebrospinal fluid

This fluid is found in the ventricular system of the brain, the central canal of the spinal cord and the subarachnoid space of the brain and spinal cord. It is produced by the *choroid plexuses* of the ventricles, especially those of the lateral ventricles (Fig.129). A choroid plexus consists of a projection of the vascular pia mater into the ventricle and it is covered by the lining cells of the ventricle. The fluid produced in the lateral ventricles (one in each cerebral hemisphere)

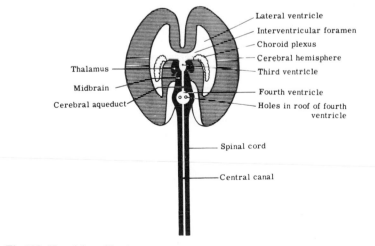

Fig.129. Ventricles of brain.

passes through the interventricular foramina (one on each side) into
the third ventricle which lies between the thalami. Some more fluid
is added by the choroid plexus of the third ventricle and the fluid
then passes through the cerebral aqueduct of the midbrain into the
fourth ventricle of the hindbrain where some more fluid is added by
its choroid plexus. A little of the fluid passes down the central
canal of the spinal cord. Most of it leaves the fourth ventricle
through foramina in its roof and thus reaches the subarachnoid
space. The fluid then spreads over the brain and the spinal cord.

It is re-absorbed into the venous sinuses through projections of
the arachnoid into these sinuses, mainly the superior sagittal
(Fig.127b). These projections are called *arachnoid villi*. Most of these
are microscopic but if they form large masses they are called *arachnoid granulations.*

Cerebrospinal fluid consists mainly of water. It contains few, if
any, cells and very little protein. It has about half the amount of sugar
and as much sodium chloride as plasma. These are the main constituents which may be changed in diseases of the meninges and central
nervous system. The fluid is under pressure which is relatively
low—*120 mm of water.* The main function of the fluid is to act as
a water cushion. It allows changes in the volume of the contents of the
cranium (these are mainly changes in the volume of the blood in the
brain).

The blood supply of the brain and spinal cord

The spinal cord receives its blood supply from many arteries which
enter the vertebral canal through the intervertebral foramina and
from descending branches of the vertebral artery, a branch of the
subclavian.

The brain is supplied by the two vertebral and two internal carotid
arteries. Both vertebral arteries enter the skull through the foramen
magnum and each internal carotid through a canal in the temporal
bone. Inside the skull the vertebral arteries join to form the *basilar
artery* which passes upwards in a longitudinal groove along the
middle of the pons. The basilar artery then divides into the right and
left *posterior cerebral arteries* (Fig.130a). This artery runs laterally
round the cerebral peduncle of the midbrain. It supplies the posterior
part of the hemisphere on all its surfaces—lateral, medial and inferior.

The internal carotid artery runs forwards and then upwards beside

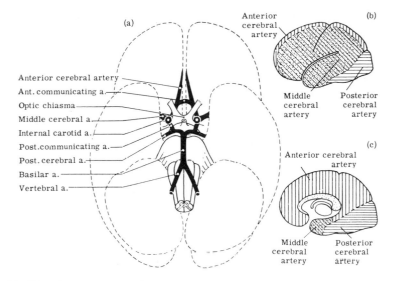

Fig.130. (a) Arrangement of large arteries supplying brain, (b) Distribution of cerebral arteries to lateral surface of cerebral hemisphere, (c) Distribution of cerebral arteries to medial surface of hemisphere.

the sella turcica of the sphenoid bone and divides, lateral to the optic chiasma, into the *middle* and *anterior cerebral arteries*. The middle cerebral passes laterally into the lateral sulcus and the anterior runs forwards for a short distance and then backwards over the upper surface of the corpus callosum between the two cerebral hemispheres (Fig.130a). The middle cerebral artery supplies the lateral surface of the hemisphere and the anterior, the medial surface (Fig.130b,c). All six cerebral arteries give off branches to the cerebral cortex and to the inside of the hemisphere. One of the branches of the middle cerebral artery is called the *artery of cerebral haemorrhage* because it is one of the most frequently ruptured vessels within the cranium. It supplies the internal capsule.

There is on each side an artery running between the posterior and internal carotid arteries (*posterior communicating*) and an artery joining the two anterior cerebral arteries (*anterior communicating*). These communicating arteries complete a ring of vessels in the region of the interpeduncular fossa called the *circle of Willis* (*arterial circle*) (Fig.130a). By means of these communications a defective blood supply due to the narrowing of one artery can be compensated for by a supply of blood from another. It is important to realize

that the arteries which enter the brain itself are end arteries, that is
they have no connexions with other arteries so that the closure of an
artery entering the brain results in the death (necrosis) of the part
supplied by that artery.

The venous blood from the surface of the brain enters the superior
sagittal sinus which forms the right transverse sinus. The blood
from the surface of the brain also drains into the inferior sagittal
sinus which joins the great cerebral vein draining the blood from the
interior of the brain. These two vessels form the straight sinus
which forms the left transverse sinus. The transverse sinus on each
side turns downwards towards the jugular foramen through which
it exits from the skull and becomes the internal jugular vein.

The main sensory and motor pathways

It is convenient to revise these pathways after completing a study of
the whole central nervous system. The main sensory pathways are
described on pp. 181 to 185. The pathways of the special senses will be
given with the cranial nerves (pp. 203 to 211).

The motor pathways are divided into pyramidal and extrapyramidal. The pyramidal fibres (or upper motor neurone) have their
cell bodies in one of the motor areas of the cerebral cortex (the
precentral gyrus or the neighbouring motor area). Their fibres pass
into the white matter of the hemisphere and then into the genu and
posterior limb of the internal capsule where they are arranged in a
specific order. From the internal capsule the fibres pass into the
cerebral peduncle of the midbrain. They then pass through the pons
and emerge from its lower border to form the pyramid of the medulla
oblongata. In the lower part of the medulla four-fifths of the fibres
cross the midline in the *decussation of the pyramids* and form the
lateral corticospinal tract. These fibres end in relation to the anterior
horn cells (lower motor neurones).

The remaining one-fifth continues on the same side as the anterior
corticospinal tract which, however, crosses over before ending in
relation to the anterior horn cells. Some of the pyramidal fibres
cross over in the midbrain and hindbrain and end in relation to the
motor cranial nerve nuclei. It can thus be seen that, broadly speaking,
the left motor cortical areas control the right side of the body.
Some muscles, however, are controlled by both sides of the brain,
for example, the muscles of respiration.

Interruption of the upper motor neurone produces a spastic

paralysis, that is movements are lost, but the muscles are in spasm. Tendon reflexes are increased, there is no wasting of the affected muscles and stroking the sole of the foot results in the big toe being bent upwards (dorsiflexion) instead of downwards (plantar flexion). Interruption of the lower motor neurone produces a flaccid paralysis. The lower motor neurone has already been dealt with. It has a cell body in the anterior horn of grey matter of the spinal cord and a fibre which leaves the spinal cord in the anterior root and is distributed to striated muscle. It supplies a variable number of muscle fibres (10 to 2,000) depending often on the size of the muscle.

Extrapyramidal is a term used to describe parts of the central nervous system which are motor in function but are not pyramidal. Included in this term are such structures as the caudate and lentiform nuclei of the forebrain, and the red nucleus of the midbrain. The term is also used to describe those tracts in the spinal cord apart from the corticospinal tracts whose fibres end in relation to the anterior horn cells. The extrapyramidal system appears to exercise some control over the pyramidal system. Disease of the extrapyramidal nuclei and/or tracts results in tremor, weakness and rigidity of the voluntary muscles and also sudden, involuntary movements (p. 196).

The cranial and spinal nerves

THE CRANIAL NERVES. There are twelve pairs of cranial nerves. Many of them are similar to spinal nerves in that they are motor and sensory to the muscles and skin of the head, but, unlike the spinal nerves, the motor and sensory nerves frequently remain separate. It is sometimes more useful to think of the various structures in the head and the neck and try to relate their nerve supply to the cranial nerves instead of the other way round. For example, the muscles of the eyeball are supplied by the third, fourth and sixth cranial nerves, the muscles of the face by the seventh cranial nerve and the muscles of the tongue by the twelfth cranial nerve. The first nerve is associated with the sense of smell, the second with the sense of sight and the eighth with the sense of hearing. Taste is associated with the seventh and ninth nerves. The skin of the head and the lining of the various cavities (mouth, nose, orbit) have a sensory innervation through the fifth nerve which also contains the motor nerve of the muscles of mastication. The palate, larynx and pharynx have muscles which are innervated by the eleventh cranial nerve.

Table 6 The cranial nerves

No.	Name	Origin	Type	Enters or leaves skull through	Functions
1	olfactory	mucous membrane of nasal cavity	sensory	roof of nose	smell
2	optic	retina of eyeball	sensory	optic canal (foramen)	vision
3	oculomotor	midbrain	motor and parasympathetic	superior orbital fissure	motor to four muscles moving eyeball; motor to ciliary and sphincter pupillae muscles
4	trochlear	midbrain	motor	superior orbital fissure	motor to superior oblique muscle of eyeball
5	trigeminal	pons			main sensory nerve to skin and mucous membrane of head (face, scalp, nasal cavity, mouth, palate); motor to muscles of mastication
	a. ophthalmic		sensory	superior orbital fissure	
	b. maxillary		sensory	base of skull	
	c. mandibular		mixed	base of skull	
6	abducent	pons	motor	superior orbital fissure	motor to lateral rectus muscle of eyeball
7	facial	pons	motor, taste and parasympathetic	temporal bone	motor to muscles of facial expression; taste from front of tongue; motor to salivary glands
8	vestibulo-cochlear (auditory)	pons	sensory	temporal bone	a. balance, position of head b. hearing
	a. vestibular				
	b. cochlear				
9	glossopharyngeal	medulla	sensory and parasympathetic	jugular foramen	sensory including taste from back of tongue and pharynx; motor to parotid salivary gland
10	vagus	medulla	parasympathetic	jugular foramen	motor to muscle of heart, lungs, alimentary tract; motor to glands of alimentary tract
11	accessory	medulla	motor	jugular foramen	motor to laryngeal, pharyngeal and palatal muscles; motor to sternocleidomastoid and trapezius muscles
12	hypoglossal	medulla	motor	occipital bone	motor to muscles of tongue

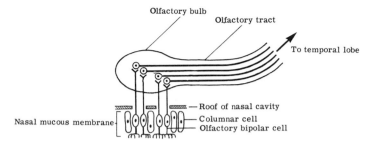

Fig.131. Olfactory nerve.

Table 6 gives a summary of the cranial nerves and this is followed by a slightly more detailed description with some diagrams which should help the student to understand their functions as well as their distribution.

The *olfactory* (1st) nerve consists of the central processes of bipolar neurones whose cell bodies are in the mucous membrane of the roof of the nasal cavity. The peripheral processes pass between the columnar cells forming the epithelium of the roof (Fig.131). The central processes form bundles of fibres which enter the cranial cavity through the roof of the nasal cavity. These processes end in the *olfactory bulb* where they synapse with the cell bodies of neurones whose fibres form the *olfactory tract*. This tract ends to some extent in the medial part of the temporal pole. This area of cerebral cortex is regarded as the smell area although the cells there and fibres in the olfactory tract have extensive ramifications. The olfactory nerve, bulb and tract, and the other structures referred to, constitute the *rhinencephalon* or *smell brain*. In man this is of very much less importance than in animals which rely on their sense of smell for information about their environment. Not only is the smell brain relatively reduced in size, but parts of it, beyond the smell area on the temporal pole, appear to have acquired new functions.

The *optic* (2nd) nerve consists of the central processes of the *ganglion cells* of the retina of the eyeball. The nerve emerges from the back of the eyeball, runs backwards and enters the cranial cavity through the optic canal in the sphenoid bone. The two optic nerves meet in the *optic chiasma* and diverge again as the *optic tracts* (Fig.132). In the chiasma the fibres from the medial halves of the retina cross but the fibres from the lateral halves remain on the same side. Each optic tract winds round a cerebral peduncle and goes to a special part of the thalamus where most of the fibres end.

No. 2

occulomotor nerve
no. 3

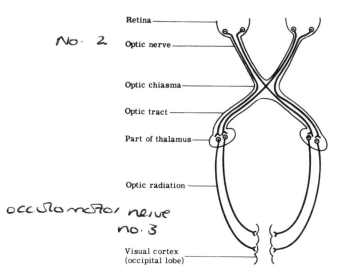

Fig 132. Optic nerves, tracts and radiations.

The fibres of the neurones with which they synapse form the *optic radiation* which passes backwards in the posterior limb of the internal capsule and ends in the visual cortex on either side of the calcarine sulcus on the medial side of the occipital lobe. Some of the fibres in the optic tract do not go to the thalamus and end in the midbrain or make an indirect connexion with the parasympathetic part of the oculomotor nerve. This is the basis of the *light reflex* in which a light shone in the eye results in a constriction of the pupil. It should be pointed out that light entering the eye acts as a stimulus to structures in the retina called *rods and cones* which in turn stimulate *bipolar cells* which stimulate the ganglion cells.

The *oculomotor* (3rd), *trochlear* (4th) and *abducent* (6th) nerves may be taken together. The abducent nerve emerges from the hindbrain at the lower border of the pons and the trochlear and the oculomotor from the midbrain. They all run forwards into the orbit through the superior orbital fissure.

The abducent supplies the lateral rectus muscle of the eyeball, the trochlear supplies the superior oblique muscle, and the oculomotor supplies the rest of the muscles which move the eyeball. The oculomotor nerve also has a parasympathetic branch. This supplies the ciliary muscle (responsible for altering the shape of the lens) and the sphincter pupillae muscle (responsible for making the pupil smaller).

The *trigeminal* (5th) nerve is mainly sensory and enters the brain at the lateral part of the pons. It also has a small motor part. The sensory part arises from the trigeminal ganglion into which three nerves pass – the *ophthalmic, maxillary* and *mandibular*. The motor part joins the mandibular nerve and supplies the muscles of mastication.

The ophthalmic nerve enters the orbit through the superior orbital fissure and is sensory to the front part of the skin of the scalp, the skin of the upper eyelid and the side of the nose (Fig.133), the structures of the eyeball and the lining of some of the air sinuses round the nasal cavity.

The maxillary nerve leaves the skull through its base, runs forward along the floor of the orbit and emerges on the face just below the lower margin of the orbit. It is sensory and supplies the skin of the side of the scalp, part of the face and lower eyelid (Fig.133), part of the inside of the cheek, the mucous membrane of the nasal cavity, palate and maxillary sinus, and the upper teeth and gum.

The mandibular nerve leaves the skull through the base. The sensory part of the nerve supplies the mucous membrane of the floor of the mouth and the anterior two-thirds of the tongue, the lower teeth and gum, and the skin of the middle of the lateral part of the scalp and part of the face (Fig.133), and part of the inside of the cheek.

The trigeminal nerve is thus seen to be the sensory nerve of the face, most of the scalp, the teeth, the mouth and the nasal cavity. It is the nerve of general sensation (pain, temperature and touch) of the skin and mucous membrane of the head.

The *facial* (7th) nerve is mainly motor and emerges from the hind-

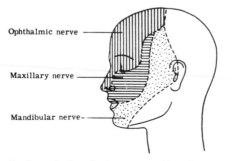

Fig.133. Distribution of trigeminal nerve to skin of scalp and face.

brain between the pons and the medulla. It has a large motor and small sensory branch. The nerve passes through the temporal bone. As it does so it passes through the internal ear, backwards between the internal ear and middle ear and then downwards to emerge from the skull from the temporal bone. It then runs forwards through the parotid gland in which it divides into several branches. These supply the muscles of facial expression. It has a branch which contains the fibres of taste from the anterior part of the tongue and parasympathetic fibres which are motor to the submandibular and sublingual salivary glands.

The *vestibulocochlear* (*auditory*, 8th) is a sensory nerve and consists of two parts each with very different functions. The vestibular nerve enters the brain at the lower border of the pons laterally. These

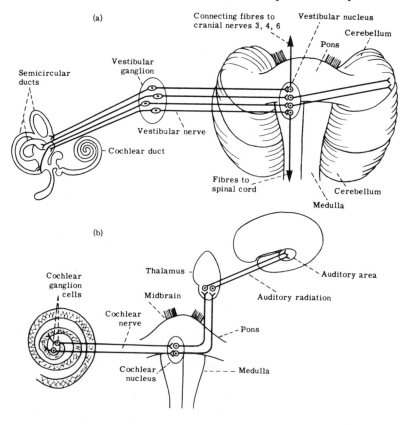

Fig.134. (a) Vestibular nerve and its connexions in brain, (b) Cochlear nerve and its connexions in brain.

fibres are the central processes of the bipolar neurones whose cell bodies are in the vestibular ganglion. The peripheral processes end in relation to structures in the internal ear. The central processes of these neurones end in relation to the vestibular nuclei (Fig.134a). The fibres of the neurones whose cell bodies form these nuclei go to the cerebellum and also send fibres down to the spinal cord forming part of the extrapyramidal system. They act on the anterior horn cells. Some fibres run upwards through the pons and the midbrain and link the internal ear with the motor nerves supplying the muscles which move the eyeball. The result is that movements of the head which stimulate the vestibular nerve affect the movements of the eyeballs.

The cochlear nerve enters the brain at the lower border of the pons. The cochlear nerve consists of the central processes of the first neurones of the auditory pathway. The peripheral processes of these neurones end in the cochlea of the internal ear and their cell bodies are in the cochlear ganglion in the internal ear (Fig.134b). The fibres of the cochlear nerve end in the pons in the cochlear nuclei and synapse there with the cell bodies of the second neurone. The axons of the second neurones cross the midline and ascend through the pons and midbrain. They end mainly in the thalamus. Some of the fibres end in the inferior colliculus. The cell bodies of the third neurones are in the thalamus and their fibres pass through the posterior limb of the internal capsule and end in the *auditory cortex* situated in the temporal lobe. Some of the fibres of the second neurones do not cross the midline so that each auditory area of the cortex, although receiving the auditory pathway mainly from the opposite ear, also receives to some extent the pathway from the ear of the same side.

The *glossopharyngeal* (9th) nerve is mainly sensory and enters the brain at the medulla. It enters the skull through the jugular foramen. As its name suggests it is distributed mainly to the tongue and pharynx. It is sensory to both these structures. In the case of the tongue it contains the fibres of general sensation and taste from its posterior part. This nerve is the sensory nerve of almost the whole of the lining of the pharynx. It has a motor parasympathetic part to the parotid salivary gland. It also has a sensory branch to the carotid sinus at the bifurcation of the common carotid artery.

The *vagus* (10th) nerve is mainly motor and emerges from the medulla. It leaves the skull through the jugular foramen and runs vertically downwards through the neck with the internal jugular vein and carotid arteries. It enters the thorax and passes to a plexus behind the hilum of the lung. It then re-forms and goes to the oesophagus round which it forms a second plexus. It emerges from this plexus and enters the abdominal cavity through the oesophagal opening in the diaphragm. In the abdominal cavity it supplies the alimentary tract as far as the middle of the transverse colon. The vagus is the main nerve of the parasympathetic part of the autonomic nervous system. The branches of the vagus supply the heart (slowing the heart), the lungs (contraction of the smooth muscle), the smooth muscle of the alimentary tract (causing contraction of the longitudinal muscle and relaxation of the circular) and secretory fibres to the alimentary glands. In the neck it also supplies some motor fibres to the muscles of the larynx, pharynx and palate although most of these are from that part of the accessory nerve which joins the vagus (Fig.135). The vagus also contains sensory fibres from the larynx, heart, lungs and alimentary tract.

The *accessory* (11th) nerve is a motor nerve and is formed by the union of the *spinal* and *cranial* accessory nerves. The spinal part comes from the upper part of the spinal cord, and passes upwards through the foramen magnum. In the cranial cavity it joins the cranial accessory nerve which emerges from the medulla (Fig.135).

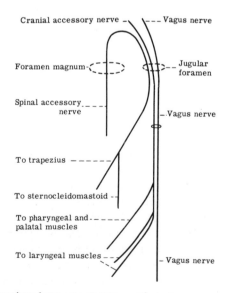

Fig.135. Connexions between accessory and vagus nerves.

The accessory nerve leaves the skull through the jugular foramen. Just below the foramen the cranial part joins the vagus and is distributed through its branches to the muscles of the larynx, pharynx and palate. The spinal part of the accessory nerve is motor to the sternocleidomastoid and trapezius muscles.

The *hypoglossal* (12th) nerve is a motor nerve and emerges from the medulla. It leaves the skull through the hypoglossal canal in the occipital bone. It supplies the muscles of the tongue.

THE SPINAL NERVES. There are thirty-one pairs of spinal nerves corresponding more or less with the bones of the vertebral column. There are eight cervical nerves instead of seven because the first emerges between the occipital bone and atlas and the eighth below the seventh cervical vertebra. There are twelve thoracic nerves, five lumbar, five sacral and one coccygeal.

The spinal nerves are formed by the union of motor and sensory roots from the spinal cord (Fig.20). These spinal nerves emerge between the vertebrae and almost immediately give off a small posterior branch. The main anterior branches except for the thoracic nerves join and divide in a very regular manner. These unions and divisions are called *plexuses* and are present on each side of the body.

The cervical spinal nerves. The upper cervical nerves (C1–C4) form the *cervical plexus* which lies in front of the upper cervical vertebrae. The branches of this plexus supply the skin of the neck and part of the ear and scalp, and many of the muscles of the front of the neck. The most important branch is the *phrenic nerve* which comes mainly from the fourth cervical nerve, passes down through the neck and thorax and supplies the diaphragm. This muscle receives its nerve supply from the upper part of the neck because it develops in a structure which in the embryo is in the neck. The diaphragm is subsequently pushed down by the developing heart and lungs and takes its nerve supply with it.

The lower cervical nerves and the main part of the first thoracic nerve (C5–C8, T1) form the *brachial plexus* in the lower part of the neck. The nerve trunks thus formed divide and rejoin as they pass into the armpit (Figs.136a,b). The final branches supply most of the skin and muscles of the upper limb. Many of the muscles round the shoulder joint are supplied by nerves which have their origin from the upper parts of the brachial plexus. The most important branches of the plexus are the following (Figs.136a,b). The *axillary (circum-*

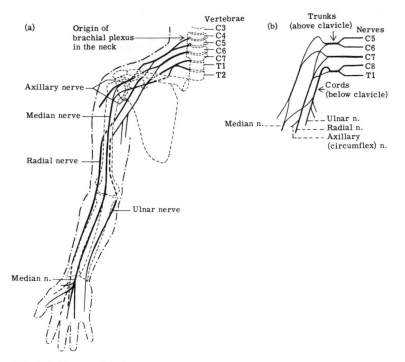

Fig.136. (a) Brachial plexus and its distribution in upper limb, (b) Formation of brachial plexus.

flex) nerve supplies the deltoid muscle, the main muscle involved in raising the upper limb sideways from the body. This nerve passes backwards below the shoulder joint and may be injured in dislocation of that joint because the head of the humerus usually dislocates downwards. The result is a paralysis of the deltoid muscle and an inability to raise the upper limb sideways.

The *radial nerve* passes behind the middle of the humerus and then downwards towards the elbow into the forearm (Fig.136a). It has a very large branch to the back of the forearm. The radial nerve supplies all the muscles on the back of the upper limb including the triceps muscle which straightens the forearm on the upper arm and the muscles which straighten the flexed hand and digits. This nerve also supplies much of the skin on the back of the upper limb including that of the hand. The radial nerve is most frequently injured near the humerus after it has given off its nerve supply to the triceps. The consequence is what is known as a *wrist drop* in which the patient finds that he cannot bend back his hand at the wrist nor straighten his bent fingers.

The *median nerve* passes down the front of the upper arm, in front of the elbow joint and into the forearm where it supplies most of the muscles which flex the hand and the fingers (Fig.136a). It then passes deep to the flexor retinaculum into the hand where it supplies some of the muscles of the thumb and the skin of the lateral three and a half digits. The median nerve supplies the muscles responsible for the movement called opposition in which the tip of the thumb is brought into contact with the tip or tips of the other digits. It also has a very important sensory supply to the tips of the thumb and index fingers. This nerve can be pressed upon as it passes deep to the flexor retinaculum in which case the patient complains of tingling and numbness in the digits supplied by the nerve. It may be cut just above the wrist. The loss of the movement of opposition and sensation in the thumb and index finger is a very marked disability.

The *ulnar nerve* passes down the upper arm to about its middle and then goes backwards so that lower down it lies behind the medial part of the lower end of the humerus where it is subcutaneous (Fig.136a). If the nerve is struck at this level there is a feeling of tingling in the little and ring fingers. This indicates its sensory distribution to the medial part of the hand and digits. At the elbow the ulnar nerve passes forwards and runs down the medial side of the front of the forearm and then into the hand where it supplies a large number of the small muscles in the hand itself, usually about fifteen of the twenty muscles.

The ulnar nerve may be injured where it lies behind the elbow. The result is that the patient has difficulty in using his hand for precise movements involving the digits. If the median nerve is injured at the elbow strong gripping by the digits is impossible.

The thoracic spinal nerves. The anterior branches of these nerves form the *intercostal nerves* which run round the thoracic wall between the ribs. They supply the intercostal muscles and the skin of the thorax. The lower intercostal nerves enter the abdominal wall and supply the muscles of the anterior abdominal wall and the skin of the abdomen.

The lumbar spinal nerves. These form a plexus called the *lumbar plexus* (Fig.137a). It has two main branches. The *femoral nerve* passes round the pelvic wall and enters the front of the thigh. It supplies the skin over the front of the thigh and the muscles which flex the thigh at the hip and straighten the leg on the thigh. The

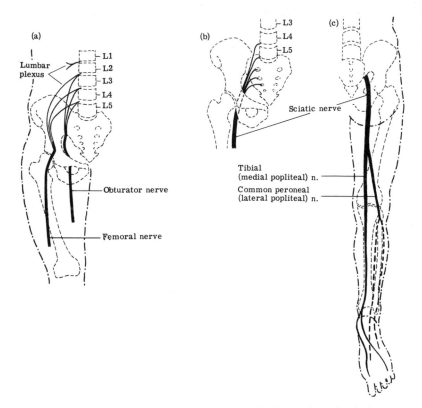

Fig.137. (a) Main branches of lumbar plexus, (b) Formation of sciatic nerve, (c) Distribution of sciatic nerve in lower limb.

obturator nerve passes round the pelvis just below its upper edge and enters the inner side of the thigh through the obturator foramen. It gives branches to the skin of the inner side of the thigh and to the hip and knee joints. It also supplies the adductor muscles of the thigh.

The sacral spinal nerves. These form the *sacral plexus* which receives a large branch from the fourth and fifth lumbar nerves. The biggest nerve of the body, the *sciatic nerve*, comes from five of these spinal nerves and enters the buttock (Figs.137b,c). It runs down the back of the thigh towards the back of the knee and usually divides about the middle of the thigh into the *tibial (medial popliteal) nerve* and *common peroneal (lateral popliteal) nerve*. The tibial nerve continues down the back of the leg, winds round the inner side of the ankle and enters the sole of the foot. The common peroneal nerve winds round the outer side of the fibula joint below the knee joint and enters the front of the leg. It runs down into the foot.

The sciatic nerve itself gives branches to the hamstring muscles which bend the leg at the knee. The tibial nerve supplies the muscles on the back of the leg (the muscles bend the foot and toes downwards) and the small muscles in the sole of the foot. It also supplies a part of the skin of the leg and foot. The common peroneal nerve gives branches to the skin and muscles on the outer side and the front of the leg and foot. These muscles turn the foot upwards and also outwards. This nerve is relatively easily injured where it lies under the skin on the fibula. The result is a *dropped foot* because the foot cannot be turned upwards. Patients with a dropped foot have to lift the affected lower limb higher than usual while walking because of this inability to turn the foot upwards. Otherwise they would scrape their toe on the ground.

There are many branches of the sacral plexus which supply the muscles and skin of the buttock. The skin of the external genitalia is supplied by a branch of the sacral plexus.

The autonomic nervous system

This part of the nervous system supplies smooth muscle and glands with motor nerve fibres. It therefore forms the nerve supply of all organs and structures with smooth muscle, for example the lungs, the alimentary tract and the blood vessels. Cardiac muscle is also supplied by the autonomic nervous system. Although mainly motor there is also a sensory part of this system but this is arranged in exactly the same way as the sensory part of the nervous system for the rest of the body. For example there are sensory fibres supplying the heart and their cell bodies are in a posterior root ganglion. The central process goes to the spinal cord in the posterior root. It is assumed that they travel in the central nervous system with the sensory fibres of the rest of the body.

Although some parts of the central nervous system are known to have autonomic functions (for example, the hypothalamus) one can regard the autonomic nervous system as being peripheral. The fibres may form separate nerves but they frequently run with the cranial or spinal nerves. There are usually two neurones involved, the first with its cell body in the hindbrain or spinal cord and the second with its cell body in an autonomic ganglion outside the central nervous system. The first neurone synapses with the second in a ganglion and the fibre of the second neurone supplies the smooth muscle of an organ, etc. The first neurone is called *preganglionic* and the second *postganglionic*.

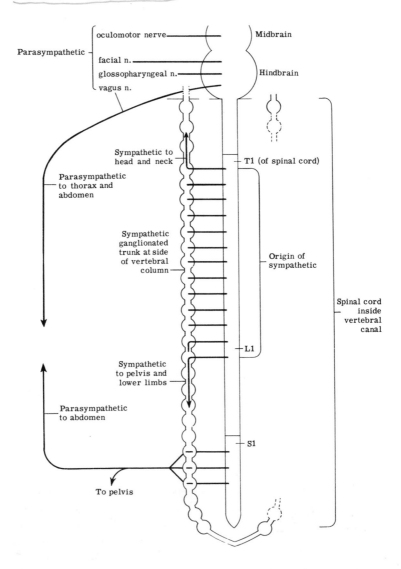

Fig. 138. Formation and distribution of sympathetic and parasympathetic nerves.

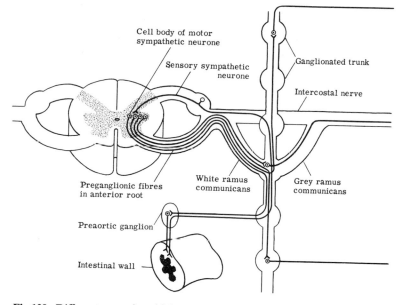

Fig.139. Different ways in which sympathetic nerves synapse in sympathetic ganglia. Anses from Arises from brain lumber and thoracic and sacral region of regions of spinal cord spinal cord

THE SYMPATHETIC AND PARASYMPATHETIC. The autonomic system is divided into the *sympathetic* and *parasympathetic*. There are certain basic differences between them.

a. If an organ has both a sympathetic and parasympathetic nerve supply they produce opposite effects; for example, the sympathetic to the eyeball causes dilatation of the pupil and the parasympathetic causes constriction of the pupil; the sympathetic causes the longitudinal muscle of the alimentary tract to relax and the circular muscle to contract and the parasympathetic does the opposite; the sympathetic increases the heart rate, the parasympathetic decreases it.

b. Ganglia of the sympathetic nerves are some distance from the organs they innervate; ganglia of the parasympathetic are near or in the organs innervated. The sympathetic ganglia form a chain extending from the base of the skull to the sacrum and also are found in front of the abdominal aorta (Figs.138, 139).

c. the terminations of the second neurone of the sympathetic produce *adrenaline* and *noradrenaline* and those of the parasympathetic produce *acetyl choline*.

d. The cell bodies of the first neurones of the sympathetic part are in the thoracic and first lumbar segments of the spinal cord and the fibres of the first neurones of the parasympathetic are in the oculomotor, facial, glossopharyngeal and vagus nerves and the second, third and fourth sacral nerves (Fig.138). With this arrangement of the origin of the autonomic nerves it is necessary for the sympathetic nerves to run upwards for the supply of the head, neck and upper limb and downwards for the pelvis and lower limbs (Fig.138). By means of the vagus nerves, parasympathetic fibres pass downwards to the thorax and abdomen, and through the sacral nerves parasympathetic fibres run upwards from the pelvis into the abdomen.

THE FUNCTIONS OF THE AUTONOMIC NERVOUS SYSTEM. The sympathetic nerve supply has often been related to what is called the preparation for fight or flight by an animal. The activity of the sympathetic results in an increase in the heart rate, relaxation of the muscle of the lung, an increase in the blood supply to the muscles and heart and a decrease in the blood supply of the skin and alimentary tract, an increase in the blood sugar and dilatation of the pupils. In addition the hairs stand on end and the sphincters (circular muscle) of the bowel and bladder are closed. The parasympathetic nerve supply, on the other hand, is said to protect organs from overactivity. It slows the heart, cuts down the light reaching the retina and improves digestion and absorption.

One can also consider the effects of the sympathetic and parasympathetic nerves on individual parts of the body. The sympathetic causes

a. the dilator pupillae to contract so that the pupil becomes bigger,

b. the blood vessels to constrict except the blood vessels of the heart and striated muscle which relax,

c. the sweat glands to secrete,

d. the hairs to stand on end,

e. the muscle of the lungs to relax,

f. the heart rate to increase,

g. the longitudinal muscle of the alimentary tract to relax and the circular muscle to contract, *h.* the glycogen of the liver to be mobilized,

i. the suprarenal medulla to secrete adrenaline and noradrenaline,

The parasympathetic causes

a. the ciliary muscle to contract with the result that the lens becomes more biconvex,

b. the sphincter pupillae to contract so that the pupil becomes smaller,

c. the lacrimal and salivary glands to secrete,

d. the muscle of the lungs to contract,

e. the heart rate to decrease,

f. the longitudinal muscle of the alimentary tract to contract and the circular muscle to relax,

g. the glands of the alimentary tract to secrete,

h. the urinary bladder to contract.

The importance of the autonomic nervous system in the normal functioning of the body can now be appreciated. By constricting and dilating the blood vessels of the skin the loss of heat by the body is to some extent regulated. Sweating is also controlled by the sympathetic nerves of the skin. The temperature-regulating centres of the hypothalamus control the activity of the sympathetic in relation to body temperature. Blood pressure is also raised and lowered by the action of the sympathetic on the blood vessels, mainly the arterioles. Various drugs which are commonly used in medicine act on the autonomic nervous system either locally, such as the use of an antiparasympathetic drug in the eye in order to dilate the pupil, and adrenaline applied to the nasal lining for nose-bleeding, or generally, such as adrenaline administered for an attack of asthma in which the muscle of the lung is in spasm. Operations are carried out in order to remove the sympathetic nerve supply of blood vessels so that they will dilate. The parasympathetic nerve supply to the glands of the stomach may be interrupted so that glandular secretion is reduced.

12

The eye and the ear

The eye

The eyeball lies in the bony orbit, is almost spherical in shape and consists mainly of three coats—the *sclera*, the *choroid* and the *retina*.

THE SCLERA. The outer coat (the *white* of the eye) called the *sclera*, consists of dense fibrous tissue and is about 1 mm thick. It is protective and contains the contents of the eyeball, which are under pressure. In front, the sclera is transparent and forms the *cornea* which is about one-sixth of the sphere formed by the eyeball (Fig.140a). The cornea has a different curvature from that of the sclera and is a segment of a smaller sphere. The most posterior part of the sclera is pierced in one area by the emerging fibres of the optic nerve. The deepest layer of the eyelids (*conjunctiva*) is reflected on to the front of the sclera. The conjunctiva is modified skin which is further modified over the cornea so that it permits light to pass through. The muscles which move the eyeball are attached to the sclera.

THE CHOROID. The middle coat is vascular and pigmented, and it is called the *choroid* where it lines the major part of the sclera. In front where it is separated from the sclera it is called the *iris*, which is the coloured part of the eye (Fig. 140a). The hole in the centre of the iris is called the *pupil*. The *ciliary body* is the part of the middle coat between the choroid and the iris. The choroid itself is very vascular and of a dark brown colour due to pigment granules in its cells. The blood vessels of the choroid are concerned with the metabolism of the eyeball and the pigment acts like the blackened inside of a camera, that is, it prevents the scattering of light entering the eyeball.

220

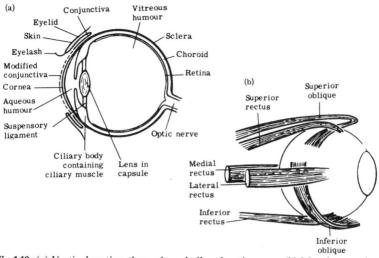

Fig.140. (a) Vertical section through eyeball and optic nerve, (b) Muscles moving eyeball.

The ciliary body is a more complex structure. It contains the *ciliary muscle* which is innervated by parasympathetic branches from the oculomotor nerve. Attached to the posterior, deeper part of the ciliary body is the *suspensory ligament,* which is attached to the covering (*capsule*) of the *lens,* a transparent structure lying behind the iris and pupil (Fig.140a). The iris is attached to the ciliary body peripherally and lies in front of the lens. In the iris are circular muscular fibres round the edge of the pupil (*sphincter pupillae*) and radially arranged fibres more peripherally (*dilator pupillae*). The former is supplied by parasympathetic (oculomotor) and the latter by sympathetic nerves. These muscles alter the size of the pupil so that the amount of light entering the eye can be controlled. In bright light the pupil constricts due to the contraction of the sphincter pupillae (*light reflex*).

When the ciliary muscle contracts, the suspensory ligament is pulled forwards so that the capsule (the outer covering of the lens) is relaxed and the lens becomes more biconvex. This is called the *accommodation reflex* and enables the eye to focus on near objects. When one focusses on near objects three things happen.

a. The lens becomes more biconvex.

b. The eyes converge so that double vision does not occur.

c. The pupil constricts.

The oculomotor nerve supplies all the muscles involved in this—parasympathetic fibres to the ciliary muscle and sphincter pupillae and somatic fibres to the muscle which pulls the eye inwards (the *medial rectus muscle* of each eyeball).

THE RETINA. The third and innermost layer is called the *retina*. It is the light-sensitive part of the eye and although it lines the whole of the choroid the anterior part next to the iris and ciliary body is one cell thick and does not respond to light stimuli. The retina is said to have ten layers but the light-transmitting units are

   a. the *rods* and *cones*,

   b. the *bipolar cells*,

   c. the *ganglion cells*.

They are connected with each other in that order from the outside towards the inner part of the eye. Light therefore has to pass through the retina in order to stimulate the rods and cones which in turn stimulate the bipolar cells and these in turn stimulate the ganglion

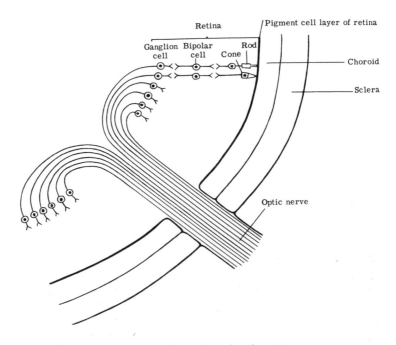

Fig.141. Basic cells of retina and formation of optic nerve.

cells (Fig.141). The fibres of the ganglion cells converge on the posterior pole of the eye and form the optic nerve. The area where they meet is called the *optic disc*. There is an outer *pigmented cell layer* of the retina, firmly adherent to the choroid.

The rods and cones react to a stimulus of light which has entered the eye through the cornea. These receptors are not distributed equally through the retina. There are no rods and cones at the optic disc which is called the *blind spot*. Lateral to the optic disc is a yellowish area (*macula lutea*) in the middle of which there are only cones. As one passes peripherally from the macula the rods become more numerous and outnumber the cones. Cones are used in bright light and for fine detail. The macula is therefore the area where the highest visual acuity is possible. Cones are also used in distinguishing colours. Rods are used in dim light but in order to see in these conditions the eye has to become *dark-adapted*. Rods contain a substance called *visual purple* which is responsible for producing the impulse transmitted by the rods. In bright light visual purple becomes bleached and in dim light it is re-formed, that is, dark-adaptation occurs. Cones do not function in this way and become only slightly dark-adapted.

THE HUMOURS (FLUIDS) OF THE EYEBALL. The *aqueous humour* is the name given to the watery fluid found between the front of the lens and its suspensory ligament, and the posterior surface of the cornea. Behind the lens and suspensory ligament the eyeball contains the *vitreous humour* which consists of a jelly-like colourless substance (Fig.140a).

REFRACTION BY THE EYEBALL. The cornea, aqueous humour, lens and vitreous humour are the refracting media of the eyeball. In a normal eye objects between infinity and 6 metres are refracted by these media and focussed sharply on the retina. If the object is nearer than 6 metres, focussing is achieved by means of changing the shape of the lens, mainly its anterior surface (the accommodation reflex). The ciliary muscle contracts so that the suspensory ligament is relaxed and the pull on the anterior part of the lens capsule is reduced. This allows the anterior surface of the lens to bulge forwards. The lens becomes more convex and near objects are focussed on the retina.

The progressive loss of the ability to accommodate after the age of 45 to 50 years is called *presbyopia*. This can be corrected by wearing glasses when viewing near objects, for example, when reading. In *myopia* (short sight) the eyeball is too large and a distant object is focussed in front of the retina. A concave lens in front of the eye focusses the object on the retina. In *hypermetropia* (long sight) the eyeball is too small and a distant object is focussed behind the retina. A convex lens in front of the eye focusses the object on the retina.

Depth and distance of vision depends on many factors. Each eye sees a slightly different image and the brain (visual cortex) fuses the two images so that stereoscopic vision results. In addition one is aware of movements of the eyes and focal adjustments in order to see objects at different distances. Other factors such as light and shade, and the interposing of nearer objects between the eyes and more distant objects, help to estimate distance.

THE MUSCLES MOVING THE EYEBALL. The eyeball is moved in the orbit by means of its *extrinsic muscles* (the muscles of the iris and ciliary body are referred to as the *intrinsic muscles*). The *superior, medial, inferior* and *lateral rectus muscles* are attached to bone in the back of the orbit round the optic canal (foramen) and superior orbital fissure, and run forwards to be attached to the sclera (Fig.140b). They form a cone which surrounds the optic nerve as it passes backwards from the eyeball. The *superior oblique muscle* is also attached to the back of the orbit and runs forwards above the medial rectus. It turns laterally in a pulley of bone (*trochlea*) and passes below the superior rectus to become attached to the sclera. The *inferior oblique muscle*, on the other hand, is attached to the bone of the front of the floor of the orbit medially and runs laterally below the inferior rectus to become attached to the sclera. The lateral rectus is supplied by the abducent nerve, the superior oblique by the trochlear nerve and the other four by the oculomotor nerve. The

Table 7

Muscle	*Pulls eyeball*	
Lateral rectus	out	
Medial rectus	in	
Superior rectus	up and in	⎫ up
Inferior oblique	up and out	⎭
Inferior rectus	down and in	⎫ down
Superior oblique	down and out	⎭

lateral rectus turns the eyeball out, the medial turns it in. The superior rectus pulls the eyeball up and in and the inferior rectus pulls it down and in. The superior oblique pulls it down and out and the inferior oblique up and out. To pull the eyeball up, the superior rectus contracts with the inferior oblique; to pull it down, the inferior rectus contracts with the superior oblique (Table 7).

The main artery of the orbit is called the *ophthalmic artery* which is a branch of the internal carotid and enters the orbit with the optic nerve through the optic canal. It gives off a branch called the *central artery of the retina* which runs in the optic nerve to the optic disc where it can be seen with an ophthalmoscope dividing into its branches. The ophthalmic division of the trigeminal is the nerve of general sensation to the eyeball.

The ear

The ear has three parts (Fig.142),

    a. the external ear,

    b. the middle ear,

    c. the internal ear.

THE EXTERNAL EAR. This consists of the somewhat oval *pinna* (*auricle*) projecting from the side of the head, and the *external*

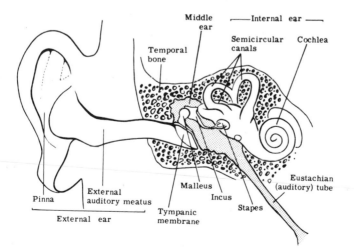

Fig.142. External, middle and internal ear.

acoustic (auditory) meatus. The auricle has several ridges and depressions which are always present to a greater or less extent. The auricle directs the sound waves towards the external meatus and is usually immobile in man although there are vestigeal muscles attached to it and the scalp. Many animals can move their ears. The auricle consists of thin elastic cartilage covered by skin.

The meatus is about 2·5 cm long and consists of an outer third which is cartilaginous and part of the auricle, and an inner two-thirds which is osseous and part of the temporal bone. It is oval in shape and curves as it passes medially both upwards and backwards so that if the auricle is pulled upwards and backwards the meatus is straightened. The narrowest part of the meatus is where the cartilaginous part joins the osseous part. The meatus is lined with skin and there are *ceruminous glands* in the cartilaginous part which produce the wax of the ear.

At the medial end of the meatus is the *tympanic membrane (ear drum)* which separates the external from the middle ear. The membrane is attached to a groove in the bone and lies obliquely so that its lower part is more medial than its upper (Fig.142). At birth it is even more horizontal and much nearer the surface. The membrane can be examined with a speculum. In its middle a projection called the *umbo* can be seen and passing upwards and forwards from it is a ridge due to the *handle of the malleus* on its inner side. Sound waves pass down the meatus and cause the membrane to vibrate.

THE MIDDLE EAR. The *middle ear (tympanic cavity)* is a space in the temporal bone. It is about 1·5 cm vertically and anteroposteriorly and about 0·5 cm from side to side. It is narrowest (about 0·2 cm) opposite the umbo because the tympanic membrane bulges inwards. The middle ear has six sides. It is important to know some of the structures related to the walls of the middle ear. A tube from the back of the upper part of the pharynx leads into the middle ear through its front wall. This tube, called the *auditory (Eustachian) tube*, is the means whereby the pressures on the two sides of the tympanic membrane are kept equal. The tube is opened when an individual swallows. A blocked tube is not an uncommon occurrence and if not relieved can lead to deafness. Infection from the pharynx can pass up the tube and lead to an infection of the middle ear. This infection can result in a perforation of the ear drum. It can spread upwards or backwards and cause a brain abscess and it can spread backwards into the mastoid process of the temporal bone. It may cause clotting in the internal jugular vein which lies just below the middle ear.

There is a chain of three auditory ossicles passing from the lateral to the medial wall of the middle ear. The *malleus* (like a hammer) is the largest and about 1 cm long. It has a head which lies in the upper part of the middle ear above the level of the tympanic membrane. Projecting downwards and backwards from the head is the handle which lies on the medial side of the tympanic membrane. The end of the handle is attached to the middle of the membrane at the umbo. The posterior surface of the head of the malleus articulates with the second bone, the *incus*, which resembles an anvil. This bone has a body which articulates with the head of the malleus. From the body there projects a long process which bends medially and articulates with the *stapes*. The stapes, resembling a stirrup, has a head which articulates with the incus. From the head two limbs diverge and are joined by a base (footpiece). The base fits into an oval hole in the medial wall. Movements of the tympanic membrane due to sound vibrations passing along the external meatus are transmitted across the middle ear by movements of the auditory ossicles in the internal ear. The vibrations of the footpiece of the stapes are transmitted to fluid in the internal ear and ultimately stimulate the fibres of the cochlear nerve.

THE INTERNAL EAR. The internal ear consists of a surrounding bony part called the *osseous labyrinth* and an inner membranous part called the *membranous labyrinth*. The osseous labyrinth is divided into an anterior part called the *cochlea*, about 1 cm long across its

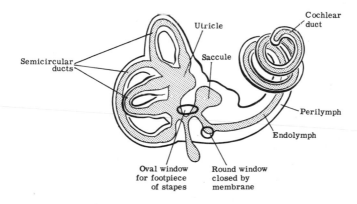

Fig.143. Membranous labyrinth inside osseous labyrinth.

base, a middle part called the *vestibule* and a posterior part called the *semicircular canals*. The cochlea is like a snail's shell lying on its side and has 2¾ turns. The vestibule, the middle part of the osseous labyrinth, has on the wall between it and the middle ear an oval opening in which lies the base of the stapes. Below this is a round opening which is closed by a membrane. There are three *semicircular canals* each about two-thirds of a circle. The *anterior* and *posterior canals* are vertical and at right angles to each other, and the *lateral canal* is horizontal. Each canal is enlarged at one end (*ampulla*).

The osseous labyrinth contains fluid called *perilymph* which resembles cerebrospinal fluid and in it lies the membranous labyrinth. This consists of an anterior part called the *duct of the cochlea* (inside the bony cochlea), two small sacs (inside the vestibule) and three *semicircular ducts* (inside the semicircular canals) (Fig.143). All these structures communicate with each other—the cochlear duct with the sacs and the sacs with the semicircular ducts. The membranous labyrinth contains fluid called the *endolymph*. The duct of the cochlea contains a complicated spiral structure which is the end organ of hearing. The fibres of the cochlear nerve end in relation to this structure.

Sound waves are transmitted by the auditory ossicles to the perilymph through the oval opening. These waves are in turn transmitted to the spiral structure inside the duct of the cochlea. This stimulates the cochlear nerve so that impulses travel along the nerve to the auditory cortex of the brain.

Parts of the semicircular ducts and parts of the sacs in the vestibule are innervated by the vestibular nerve. The areas supplied by the nerve have special structures on them. Movement of the endolymph stimulates these structures which in turn stimulate the fibres of the vestibular nerve. Forward, rotatory and side-to-side movements of the head move the endolymph and lead to stimulation of the nerve. In this way the brain is kept informed about the movements of the head (and body). This information is also conveyed to the muscles which move the eyes and the head. As has been already stated the vestibular nerve is linked to the cerebellum, a structure which co-ordinates movements of the different parts of the body.

Giddiness, which follows spinning round several times, is due to the endolymph continuing to move after the spinning has stopped. The result is a continued stimulation of the vestibular nerve giving the sensation of the outside world moving. A secondary effect is staggering due to inco-ordination of the muscles of the limbs which do not receive their normal instructions leading to co-ordinated movement. A typical symptom of vestibular disease is a complaint of spontaneous attacks of giddiness.

13

The endocrine glands

By definition these are glandular organs producing secretions which pass straight into the blood stream. Because they have no ducts they are often referred to as *ductless glands*.

The hypophysis cerebri (pituitary gland)

The *hypophysis cerebri* is about the size of a large pea and lies in the hypophyseal fossa (sella turcica) of the sphenoid bone inside the skull below the hypothalamus of the forebrain. It is attached to the floor of the third ventricle by the *infundibulum* (stalk). The hypophysis consists mainly of two lobes, an *anterior* which develops upwards from the roof of the stomodeum (primitive mouth) and a *posterior* which develops downwards from the floor of the third ventricle (Fig.144a).

The anterior lobe produces a number of hormones many of which act on other endocrine glands. These hormones include the following,

a. *Growth (somatotrophic) hormone* (GH): excess of this hormone during growth results in *gigantism* in which the individual is excessively tall (over 2·4 m). Excess in adult life produces *acromegaly* in which the hands, feet and chin are very large. Lack of this hormone during the growth period produces *dwarfism* in which the individual is small but normally proportioned.

b. *Adrenocorticotrophic hormone* (ACTH): this hormone acts on the cortex of the adrenal gland and is responsible for its production of hormones.

230

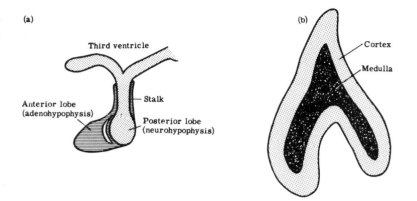

Fig.144. (a) Pituitary gland (hypophysis cerebri), (b) Adrenal gland.

c. *Thyroid stimulating hormones* (TSH): this hormone stimulates the thyroid gland to produce its hormones.

d. *Follicle stimulating hormone* (FSH): this hormone acts on the ovarian follicle and is responsible for the maturation of the ovum and follicle.

e. *Luteinizing hormone* (LH): this is responsible for the development of the corpus luteum.

f. *Prolactin*: this hormone is responsible for the secretion of milk during lactation.

g. *Melanocyte stimulating hormone* (MSH): this acts on the pigment-producing cells of the skin so that it becomes darker.

It may be added that in the male the gonadotrophic hormones (d) and (e) act on the testis. Both FSH and LH are involved in spermatogenesis and the production of testosterone by the interstitial cells of the testis.

The posterior lobe produces an *antidiuretic hormone* (ADH) which acts on the tubules of the kidney so that water is re-absorbed in sufficient quantities. It thus helps to regulate water balance. It also produces a hormone which acts on smooth muscle. This is called *oxytocin* which acts on the smooth muscle of the uterus. ADH is the same as *vasopressin* which in large doses acts on the smooth muscle of blood vessels.

The hypothalamus produces hormones which control the production of the hormones of the anterior pituitary and the hormones of the posterior pituitary are actually formed in the hypothalamus.

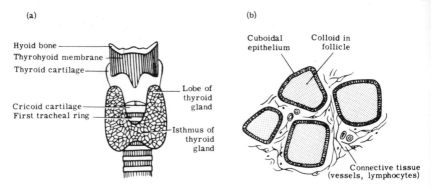

Fig.145. (a) Thyroid gland, (b) Histology of thyroid gland.

The thyroid gland

The *thyroid gland* lies in the neck and consists of two pear-shaped lobes one on either side of the lower part of the larynx and upper part of the trachea, joined by a narrow piece of thyroid tissue, called the *isthmus*, at the level of the second, third and fourth tracheal rings (Fig.145a). The thyroid gland consists of a large number of follicles containing colloid (Fig.145b). This colloid contains the hormones *thyroxin* and *tri-iodothyronine* and both contain iodine.

The effects of the thyroid hormone have been studied in clinical conditions. Generally speaking thyroxin stimulates metabolism in the tissues so that oxygen consumption and heat production are increased. In a condition known as *toxic goitre* the patient loses weight, shows tremor of the outstretched fingers, has an increased heart rate (100+ per minute), is generally excitable and sweats excessively. If this is accompanied by protrusion of the eyeballs, the condition is known as *exophthalmic goitre*. Removal of almost the whole of the thyroid gland usually results in the disappearance of these signs and symptoms.

Lack of thyroid hormone in an adult produces a condition known as *myxoedema*. The patient puts on weight, has a slow heart rate, and is mentally dull and slow, and the skin is dry and coarse. If given thyroid hormones a myxoedematous patient becomes normal.

In districts in which there is a lack of iodine in the drinking water a condition known as *simple goitre* is found. In simple goitre the thyroid gland is very enlarged because it is attempting to produce enough hormones. Apart from the swelling in the neck, the patient is fairly normal. The children of women suffering from simple goitre may be born without a thyroid in which case they are *cretins*. These children show marked arrest of growth both mentally and physically. If cretins are given thyroid hormones early enough they often become normal.

The thyroid gland also produces *calcitonin* which is important in calcium metabolism.

The parathyroid glands

The *parathyroid glands* have this name because they are found in or near the thyroid gland. In man there are four small glands about 0·5 cm long, 0·3 cm wide and 0·2 cm thick usually embedded in the upper and lower parts of the back of each lobe of the thyroid gland.

The hormone produced by the parathyroid glands regulates the amount of calcium in the plasma. Too little hormone results in a reduction of the calcium circulating in the body and this leads to hyperexcitability of the nervous system (*tetany*) in which muscles go into spasm as a result of relatively slight stimuli. Too much hormone results in a rise in the calcium of the blood and generalized weakness. If this persists long enough the bones become decalcified and spontaneous fractures may occur.

The adrenal (suprarenal) glands

The *adrenal gland* is situated in front of the upper pole of a kidney. It is about 5 cm long, 3 cm wide and 0·5 cm thick and consists of an outer *cortex* and an inner *medulla* (Fig.144b). The cortex produces several hormones including a small amount of male and female sex hormones. The main hormone is called *cortisone* (*cortisol*) and is important in water and electrolyte balance and the maintenance of blood pressure. It is also involved in carbohydrate and protein metabolism. Cortisone also prevents the normal tissue response to injury and infection. This may or may not be desirable. It also reduces the production of lymphocytes and therefore reduces the immune response of the body.

The medulla produces *noradrenaline* and *adrenaline*. Their effects are similar to those produced by stimulation of the sympathetic nerves (p. 218) and their production is said to be a preparation for fight or flight. One of the main differences between noradrenaline and adrenaline is that the former is more potent in raising the blood pressure.

The islets of Langerhans and the gonads

The *islets of Langerhans* of the pancreas and their internal secretion
insulin have already been described (p. 152). The ovary (p. 165)
produces two hormones which act on the uterus and several other
organs, and the testis (p. 163) produces the male sex hormone.

14

The skin

The skin is the outer covering of the body. It consists of two layers, an outer called the *epidermis* and an inner called the *dermis* or *corium* (Fig.5a).

The epidermis

The *epidermis* consists of stratified squamous epithelium. It has several layers but its thickness varies over different parts of the body. This variation is usually due to a variation in the most superficial layers of non-cellular *keratin* which forms the outermost layer of the epidermis. It is especially thick over some areas of the palms of the hands and soles of the feet.

The surface of the skin may be marked by deep lines, for example, the palm of the hand (*flexure lines*) or finer lines, for example, the furrows and ridges of the finger tips. The arrangement of the latter are peculiar to the individual and form the basis of identification by finger-printing. The former are used by people who profess to be able to foretell the future (palmistry) (Fig.146).

The deepest layer of the epidermis is called the *stratum basale* and all the more superficial layers develop from this. The layers have different names and their structure gradually changes as they approach the surface. The main changes involve the production of keratin and the death of the cells.

The epidermis grows down into the underlying dermis and forms *hair follicles, sebaceous glands* and *sweat glands* (Fig.147). *Hairs* grow from hair follicles. The deepest part of the hair arises from the hair bulb into which projects the dermis (papilla of the hair) (Fig.5a). Hairs vary in length and thickness and may be curly or straight. The number of hairs varies with age and sex and in different regions of the skin.

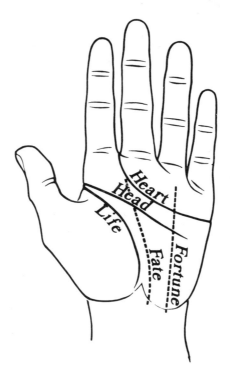

Fig.146. Flexure lines in palm of hand (used in palmistry).

Associated with the hair follicles in most areas of the skin, especially in the face and the scalp, are *sebaceous glands*. They are found in the dermis and consist of large cells containing fat. The cells produce *sebum* which is conveyed by a wide duct into the hair follicle and thence to the surface. Sebum is greasy and acts as a lubricant for the hair and skin. It also may be bactericidal and it protects the skin from being affected by water and from drying up.

Sweat glands are found in all parts of the skin and consist of a body in the dermis and a duct passing upwards and opening on to the surface of the epidermis. The body is coiled. The size and number of sweat glands vary in different parts of the body, for example, they are large in the axilla (arm pit) and numerous on the palms of the hands. Sweat glands are controlled by sympathetic nerves which also innervate bundles of muscle fibres related to the hair follicles and sebaceous glands. The muscles (*arrectores pilorum*, singular

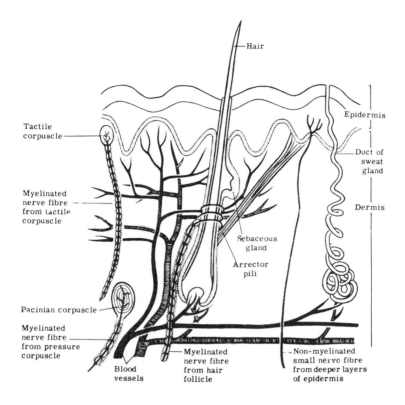

Fig.147. Details of structures in dermis of skin.

arrector pili) pass obliquely from a hair bulb below a sebaceous gland towards the deeper layers of the epidermis. When they contract they pull the hair erect, pull the skin in (goose pimples) and may squeeze the sebaceous gland. Sweat glands are important in temperature control. They can incidentally affect water balance and to some extent the excretion of waste substances.

Nails are also modifications of the epidermis (Fig.5c). They have a *root* embedded in the skin and a *body* which is the exposed part. The whiter part of the nail near the root is the *lunule* and is more opaque than the rest of the nail which looks pink due to the underlying blood vessels shining through the translucent nail. The *nail bed* is the area on which the nail lies. The nail bed consists of the stratum basale but only the proximal part of the nail bed, that is the part under the root, can form new nail substance.

The dermis (corium)

The *dermis* consists of vascular connective tissue with a large number of elastic fibres. It varies in thickness and is thin, for example, in the eyelids and thick on the soles of the feet. It also contains lymph vessels and nerves. The deeper part of the dermis is denser than the more superficial. The surface of the superficial part has projections (*papillae*) which fit into indentations on the deep surface of the epidermis. In the papillae are found capillaries and also touch corpuscles. The dermis also contains the deeper parts of the sweat glands and hair follicles and the whole of the sebaceous glands.

The functions of the skin

The skin protects the body by providing an external, relatively hard, keratinous surface which is non-cellular. The result is that slight knocks or abrasions remove only surface structures which are dispensable and readily replaced.

The skin prevents the entry into and loss of water from the body. Water is lost with the secretion of sweat but this is connected with temperature regulation. It should be noted that the kidney and not the skin is the organ controlling water balance. If there is no intake of water the body continues to sweat in order to lose heat.

Entry of harmful bacteria into the body is prevented by the skin. This may be helped by the sebum. There are many harmful organisms on the skin but only if the skin is broken do they produce inflammation. These organisms can be transferred to surgical wounds and precautions against this possibility have to be taken.

The skin is an important sensory organ. Stimuli which are interpreted as painful travel along free nerve endings which end in the epidermis and dermis. Some nerve fibres end in relation to hair follicles. Movement of the hairs stimulates these nerve fibres and the resulting impulse is interpreted as touch. There are also special end organs in the skin. Some are associated with touch (Meissner's corpuscles in the papillae of the dermis) and some with pressure (Pacinian corpuscles in the deeper parts of the dermis). There may be end organs associated with the senations of heat and cold.

The skin is of special importance in heat regulation. Heat is lost from the skin by radiation, convection and conduction and also by evaporation of water associated with insensible perspiration. The amount of heat lost by the skin depends partly on the blood flow through the skin. This is increased with increased heat production, for example, as a result of muscular exercise. The increased temperature of the circulating blood acts on the hypothalamus which in turn acts on the nerves supplying the arterioles of the skin. Vasoconstrictor impulses are *reduced* and the blood vessels dilate. This leads to increased heat loss by radiation, convection and conduction. In addition the same nerves stimulate the sweat glands and sweating increases, so that more heat is lost by evaporation. If excessive, this can lead to loss of sodium chloride and result in muscular *heat cramp*. More heat can be lost by removing clothing and by consuming cold food and fluid. The loss by convection is increased by means of fans. If the surrounding temperature is higher than that of the body or the surrounding air is saturated with water vapour, heat loss is much more difficult.

In normal conditions about five-sixths of the heat produced by the body is lost by means of the skin. Sweating, although primarily associated with heat regulation, is also associated with the excretion of some sodium chloride and a small amount of urea.

The changes in the blood flow through the skin by constriction and dilatation of the blood vessels are also associated with changes in blood pressure and the transfer of blood from one part of the body to another.

Vitamin D is also formed by the skin. Ergosterol, normally found on the epidermis, if irradiated by ultra-violet light, forms vitamin D. Vitamin D is required for the proper formation of bones and teeth and lack of vitamin D causes *rickets*, a disease which is not seen in countries where the sun shines frequently and for a long time.

The skin can also form pigment to protect the body from the harmful effects of the rays of the sun. A certain amount of pigment (*melanin*) is found in normal skin and the quantity varies in different races.

Appendix

A knowledge of anatomy and physiology is necessary to answer parts of the questions which appear in the Final State Examination of the General Nursing Council for England and Wales. The questions from papers set in the past four years involving anatomy and physiology have been grouped to correspond with the Chapters of this book but the actual questions themselves have not been used. Instead, similar questions and suggested answers have been formulated and it is hoped that these will prove to be useful to candidates for the Final State Examination.

CHAPTERS 3, 4   THE LOCOMOTOR SYSTEM

Q. Describe with the aid of a diagram the joint involved in a prolapsed lumbar intervertebral disc. (pp. 47, 48, Fig.34d)

A. The joint between the bodies of the vertebrae is a cartilaginous joint in which the two bones are united by a fibrocartiliginous intervertebral disc. The bone surfaces are covered by hyaline cartilage. The intervertebral disc has a tough fibrous outer part (*anulus fibrosus*) and a more fluid inner part (*nucleus pulposus*). When bending the trunk forwards, especially when lifting heavy weights, the tough outer part can be torn at the back (usually to one side) and the softer centre bulges out. This is called a prolapsed disc. If this occurs in the lower lumbar region (the commonest sites of a prolapsed disc are between the fourth and fifth lumbar vertebrae and between the fifth lumbar and first sacral vertebrae) the appropriate spinal nerve as it enters the intervertebral foramen can be pressed on so that backache and pain down the lower limb can ensue.

Q. Describe the hip joint by means of a labelled diagram. (p. 44, Fig.32a,b,d; p. 77)

A. The diagram is not sufficiently labelled for the purpose of the answer but one can see that the hip joint is a synovial, ball and socket joint formed by the cup shaped *acetabulum* of the hip (innominate) bone and the rounded *head* of the femur. The part of the femur next to the head is called the *neck*. The upper prominence to the right is the *greater trochanter* and the lower prominence to the left is called the *lesser trochanter*. In osteo-arthritis of the hip the articular cartilage covering the acetabulum and the head of the femur is affected and is often absent and the edges of the articular surfaces show bony outgrowths. Movements are therefore painful and limited.

Q. Describe the structure of the knee joint by means of a labelled diagram in order to explain the operation for removal of a torn cartilage (meniscus). (p. 44, Fig.32d,e,f; p. 49, Fig.35e; p. 79–81, Fig.53b)

A. The knee joint is a synovial, condyloid joint between the condyles of the femur and the condyles of the tibia. In addition the patella (knee cap) articulates with the front of the lower end of the femur. The femur and tibia are held together by a fibrous capsule and a large number of strong ligaments. The rounded condyles of the femur do not fit closely on to the flat upper surfaces of the tibial condyles. The two bones are separated by a crescentic cartilage (meniscus) on each side. The shape of these cartilages are such that they deepen the upper surface of the tibia to which they are attached. Sometimes one of the cartilages (more commonly the medial) is caught between the articulating surfaces of the femur and tibia, and is torn. It will not heal and has to be removed by an incision into the joint at the side of the patella and patellar ligament. Following removal of the torn cartilage a complete recovery of function can be expected.

CHAPTER 5   THE CARDIOVASCULAR SYSTEM

Q. In terms of anatomy and physiology, explain the signs and symptoms which occur in an elderly man suffering from chronic bronchitis and congestive cardiac failure. (pp. 89–91)
A. The patient will be breathless. He will have attacks of coughing and will produce considerable quantities of thick, purulent sputum. This is due to infection of the respiratory passages involving the bronchi and their subdivisions. There is also a great increase in the number of mucous (goblet) cells of the bronchioles, and the mucous glands of the bronchial wall are hypertrophied. The cilia also disappear.

Due to failure of the right side of the heart (p. 89, Fig.59b) there will be a rise in venous pressure and an accumulation of fluid especially in the lower limbs, with swelling (oedema). All the venous blood returns to the right atrium by the superior and inferior venae cavae. The veins of the neck may be visibly distended because of the rise in venous pressure.

The failure of the left side of the heart results in back pressure of fluid in the lungs since the blood that the left ventricle normally pumps round the body returns to the left atrium by way of the pulmonary veins (p. 89, Fig.59b). The result is an accumulation of fluid in the lungs especially their bases. This will accentuate the breathlessness and reduce the oxygenation of the blood. This, together with the venous pressure and chronic bronchitis may produce cyanosis (blueness of the skin).

Q. In terms of anatomy and physiology explain the signs and symptoms of cardiac asthma (nocturnal paroxysmal dyspnoea). (pp. 89–91)
A. Attacks of nocturnal paroxysmal dyspnoea, in which the patient wakes up during the night with a sudden attack of breathlessness, are due to failure of the left ventricle of the heart. Normally the blood returns to the left atrium from the lungs via the pulmonary veins (p. 89, Fig.59b). It passes from the left atrium to the left ventricle from which it is pumped round the body. The consequence of the failure of the left ventricle is the accumulation of fluid in the pulmonary tissues. This results in the inefficient functioning of the lungs and the possibility of breathlessness. When upright the pressure in the pulmonary vessels is less than the pressure when lying down. The spasmodic attacks of breathlessness at night are due to the patient lying down. This leads to an increase in the pressure in the vessels of the lungs and results in the nocturnal attacks of breathlessness.

The term 'asthma' is not a good one since asthma should be used to describe conditions in which there is spasm of the smooth muscle of the respiratory passages. However, in the condition described above there may be added spasm of the bronchial muscle in which case the term 'cardiac asthma' is justified.

Q. What are the main similarities and differences between the structure of arteries and veins? (pp. 95–98, Fig.63)

A. Both arteries and veins consist of three coats, an innermost called the tunica intima, a middle called the tunica media and an outer called the tunica adventitia. The tunica intima consists of a single layer of squamous epithelium and is the same in both arteries and veins. Many veins have projections of the lining epithelium and form valves consisting of two flaps (Fig.63d). These are never seen in arteries.

The tunica media in both arteries and veins consists of smooth muscle and a variable amount of elastic tissue. This coat is much thinner in veins than in arteries and also contains much less muscle and elastic tissue. Different arteries also vary much more in the structure of this coat so that there are muscular and elastic arteries.

The tunica adventitia consists of connective tissue in both arteries and veins but is thicker in veins than in arteries.

The central hole or lumen of a vein is much larger than that of an artery of comparable size.

CHAPTER 6   THE BLOOD AND THE LYMPH

Q. Describe briefly the composition of the blood. (pp. 108–112)

A. (In answering this part of a question it is important that the candidate does not spend too much time on what he/she knows about the composition of the blood.)

Blood is a fluid tissue circulating in the heart and blood vessels. It consists of a fluid part called the plasma and a cellular part (the blood cells or corpuscles) suspended in the fluid. If the blood is allowed to clot the expressed fluid is called serum. The cells constitute less than half the blood (about 45 per cent). There are about 6 litres of blood in the body.

Plasma is largely water (more than 90 per cent) and the greatest proportion of the remaining constituents are the plasma proteins (about 7 per cent). Dissolved in the plasma are large numbers of substances which may be grouped as (1) inorganic substances, for example salt (sodium chloride), iodine, iron (2) organic substances, for example urea and glucose (3) hormones and antibodies.

The cells or corpuscles of the blood are grouped under these headings (1) red blood corpuscles (erythrocytes) which have no nucleus and are the means whereby oxygen is transported round the body (2) white blood corpuscles (leucocytes) which are nucleated and are mainly concerned with defence mechanisms of the body (3) blood platelets (thrombocytes) which are non-nucleated and are essential for normal clotting of the blood. The red blood corpuscles are much more numerous than the white (roughly there are 500 red corpuscles to 1 white). The blood platelets if counted are about $\frac{1}{20}$ in number as compared with the red corpuscles.

Q. What are the structure and functions of the white blood cells (leucocytes)? (p. 110; pp. 114–116)

A. Blood consists of a fluid part called plasma (55 per cent) and cellular part (45 per cent). The cellular part is divided into (1) red blood cells (corpuscles) (2) white blood cells and (3) blood platelets. The white cells are nucleated, the other two are not.

The white cells number about 8000 per c. mm of blood and they are much fewer than both the red blood cells and platets. The white blood cells are about 7–10 μm in diameter. They are divided into cells with granules and cells without granules. (These granules can be seen by special staining methods.) Thus there are granular and non-granular white blood cells. The granular have a lobed nucleus and are therefore called polymorphonuclear leucocytes (p. 109, Fig.68) (leucocyte is another name for white blood cells) or 'polymorphs' for short. Granular leucocytes constitute 75 per cent of the white blood cells and the non-granular, called lymphocytes, for the remaining 25 per cent. About 5 per cent of the total white cells are called monocytes which are slightly larger than the granular leucocytes and have a kidney-shaped nucleus. Of the granular leucocytes the great majority are called neutrophil leucocytes. The others are divided into eosinophil and basophil leucocytes according to the staining properties of the granules.

All leucocytes are concerned with the defence mechanisms of the body. Neutrophil leucocytes are specially concerned with acute infections and in these conditions they increase greatly in number both in the blood and round an area of infection. These cells are phagocytic, that is they 'eat' organisms and other particulate matter.

Non-granular leucocytes or lymphocytes are involved in the defence of the body against many types of infection and produce antibodies which make the body immune to further infection, for example vaccination against smallpox or immunity following an attack of measles. Lymphocytes are also involved in the processes of the body which result in the rejection of a graft from another person. The monocytes are phagocytic and find their way into different tissues, such as the liver and lungs, where they engulf and destroy bacteria and may be involved in the destruction of tumour cells.

Q. Make a list of the constituents of the blood. What are the functions of the cells? (pp. 108–112)
A. Blood consists of plasma (55 per cent) and cells (45 per cent). The following is a table of their constituents.

Plasma	Cells
Water (92 per cent)	Red blood corpuscles
Plasma proteins especially albumin and globulin (7 per cent)	White blood corpuscles – granular (polymorphonuclear) and non-granular (lymphocytes)
Inorganic substances, for example, sodium, potassium, calcium, magnesium, iodine, iron, often in the form of salts, for example, sodium chloride. Organic substances, for example, glucose, amino acids, neutral fats, urea	Platelets
Hormones and antibodies	

The main function of the red blood cells (erythrocytes) is to transport oxygen from the lungs to the tissues. The white blood cells (leucocytes) are involved in the defence mechanisms of the body, for example, the granular cells in acute infections and the non-granular in immunity. The blood platelets (thrombocytes) are involved in the normal process of blood clotting.

CHAPTER 7   THE RESPIRATORY SYSTEM

Q. Describe the mechanism and physiology of respiration. (pp. 62–67; 126–130)

A. Respiration is the means whereby oxygen is taken into and carbon dioxide is expelled from the body. It has two phases, inspiration in which the lungs expand and take in air and expiration in which the lungs contract and expel air. Inspiration involves the enlargement of the thorax due to a downward movement of the diaphragm and an upward and outward movement of the ribs. Expiration involves the opposite movements and is due to relaxation of the muscles used in inspiration and the elastic recoil of the lungs.

Inspired air contains about 21 per cent oxygen and almost no carbon dioxide. When the air reaches the smallest parts of the respiratory passages (the alveoli) oxygen passes from the alveoli to the capillaries which surround the alveoli because it is under greater pressure than the pressure of the oxygen in the red cells in the capillaries. In a similar way the carbon dioxide in the blood passes from the blood to the alveoli because it is under higher pressure than the carbon dioxide in the alveoli. Not all the oxygen is taken up by the red blood corpuscles and not all the carbon dioxide is given off by the blood. The result is that expired air contains about 16 per cent oxygen and 4 per cent carbon dioxide.

This exchange of gases in the alveoli can be adjusted to the requirements of the body so that by rapid deep breathing more oxygen can be taken in and more carbon dioxide can be expelled. This is seen in a normal person who is taking exercise.

Q. Describe the trachea.
A. See pp. 124, 125, Figs.76, 84.

Q. By means of anatomy and physiology, explain to a beginner in nursing what causes an attack of bronchial asthma. (pp. 128–130)

A. The respiratory passages lead from the nose (and mouth) to the larynx, trachea and main bronchi, of which there are two, a right to the right lung and a left to the left lung. The main bronchi divide into lobar bronchi each of which goes to a lobe of the lung, three in the right lung and two in the left. These lobar bronchi divide into smaller and smaller air tubes and when quite small they are called bronchioles. These in turn divide until they end in air sacs which have on their walls small protuberances called alveoli.

From the trachea onwards the walls of the air passages contain smooth muscle as far down as the smaller brochioles. The smooth muscle, if it contracts, narrows the diameter of the air passages and within certain limits changes in diameter will take place according to body requirements. For reasons not fully understood people suffering from asthma are liable to have attacks of spasm of the smooth muscles of the respiratory passages so that they have difficulty in breathing because of the narrowing of the passages.

The smooth muscle may go into spasm because of infection or because the patient is allergic to certain substances such as feathers, pollen, etc. The smooth muscle of the lung is activated by what is called the parasympathetic part of the autonomic nervous system. The muscle is inactivated by the opposite part of this nervous system, the sympathetic. That is why the active substance of the sympathetic system, adrenaline, is given in an asthmatic attack. It is also the reason why an antiparasympathetic substance called atropine is given before an operation – the anaesthetic may cause spasm of the smooth muscle. This drug will also prevent the production of mucus by mucous glands which are activated by the parasympathetic.

CHAPTER 8 THE ALIMENTARY SYSTEM

Q. A patient has a perforated appendix. By means of anatomy and physiology (a) explain the clinical features of this condition (b) explain the postoperative complications. (pp. 148–149, Fig.90)

A. (These are two separate questions.)

(a) The appendix, more correctly the vermiform appendix, is about 9 cm long and 1 cm wide and lies in the lower right part of the abdominal cavity called the right iliac fossa (this part is now called the right inguinal region of the abdomen). The appendix is a blind ending tube and part of the large intestine. It is attached to the caecum which is that part of the large intestine below the entrance of the ileum, the terminal part of the small intestine.

The earliest stage of appendicitis involves only inflammation of its interior. This is associated with pain round the umbilicus because the appendix and that part of the abdominal wall have the same nerve supply. The patient may then have a bout of vomiting or a loose stool. Because of the inflammation the patient's temperature and pulse will be raised.

As the inflammation spreads through the wall of the appendix, its outer surface, that is its peritoneal covering, becomes inflamed and the right iliac fossa is involved. The pain is now felt in the right iliac fossa and on examination there is tenderness and muscle contraction (guarding) when the right iliac fossa is palpated.

The inflammation now weakens the wall of the appendix and it may perforate. This spreads the inflammation but it may be limited by a movable apron of peritoneum called the greater omentum. A more serious complication is a spread of the infection through the whole abdominal cavity. This is peritonitis. The patient will be very ill with a high temperature and rapid pulse. The whole abdominal wall is very tense and is described as 'board-like' when felt.

(b) After the removal of a perforated appendix although the inflammation appeared to be limited at the operation it is possible for an abscess to form in the region of the appendix or for the inflammation to spread throughout the abdominal cavity (peritonitis) (see above).

The patient complains of pain on breathing, especially deep breathing and coughing, because deep expiration and coughing require contraction of the muscles of the abdominal wall.

Another complication is a paralysis of the whole intestine, called ileus. Almost any abdominal operation involving handling of the intestine can be followed by this complication. The abdomen becomes distended and the patient passes neither faeces nor flatus.

Q. By means of anatomy and physiology explain the clinical features of pyloric stenosis. (pp. 145–146, Fig.91)
A. The pylorus is the region between the stomach and the duodenum, the first part of the small intestine. Normally the circular muscle is thickened in this region, and forms the pyloric sphincter. Pyloric stenosis occurs either as a congenital condition in which there is gross hypertrophy (overgrowth) of the sphincter, or as an acquired condition due to scarring of this region as a result of an ulcer.

Clinically the most striking feature is projectile vomiting in which the contents of the stomach are vomited with considerable force due to the accumulation of food in the stomach because the food cannot pass into the duodenum. There is distension of the abdomen especially in the upper left quadrant. There is loss of weight due to lack of food and visible movements of the stomach may be seen through the thin abdominal wall.

Q. By means of anatomy and physiology explain to a junior colleague why a patient experiences considerable discomfort following cholecystectomy and exploration of the common bile duct. (pp. 150–151, Figs.93, 94; pp. 65–66)
A. This operation involves making a large incision into the upper part of the abdominal wall just below the right costal margin. Frequently a tube is inserted into the common bile duct and emerges from the abdominal incision. This tube ensures postoperative drainage and can be used for subsequent X-ray investigations. Because of the wound, breathing is difficult and uncomfortable. All types of respiration whether shallow or deep, and especially coughing, involves movement of the abdominal wall and rib cage. The wound is near the rib cage. The main cause of the discomfort is due to the respiratory movements involving the muscles of the abdominal wall. This is also the explanation for the increased incidence of respiratory complications following operations in which the incision is made into the *upper* abdominal wall.

The operation may be followed by distension of the stomach or intestine. This would lead to discomfort. The possibility of infection and/or leaking of bile into the peritoneal cavity should be kept in mind as a possible cause of the discomfort.

Q. By means of a labelled diagram explain what is meant by a 'hernia'.
(pp. 70–71, Fig. 49; pp. 141–144, Fig.86)
A. The diagram in Fig.49 can be used for this answer. A hernia has two basic elements (a) the protrusion of the lining membrane of any cavity of the body (the commonest is the peritoneum of the abdominal cavity) (b) some of the contents of the cavity (the commonest is a loop of small intestine into an abdominal hernia).

The commonest hernia is an inguinal hernia which is due to a protrusion of peritoneum down the inguinal canal. This may be either congenital or acquired. (Everybody, male and female, has a protrusion of peritoneum along the inguinal canal during fetal life (Fig.49). This protrusion is not visible unless something, for example, a loop of small intestine enters the protrusion. That is why a hernia usually 'disappears' when the patient lies down and returns on standing up.)

A protrusion of peritoneum can also occur (a) near the umbilicus (umbilical) (b) following an operation involving an incision into the abdominal wall (incisional) (c) along a canal on the inner side of the femoral vessels just below the inguinal ligament (femoral).

CHAPTER 9   THE URINARY SYSTEM

Q. Describe the gross structure and anatomical relations of the right kidney. (pp. 155–156; Figs.99, 100)
A. The kidney has a characteristic shape (bean-shaped) and is 10 cm long, 6 cm wide and 3 cm thick. Its gross structure is seen by means of a section parallel to its anterior and posterior surfaces from the medial to the lateral border. In this section the ureter can be seen expanding into its pelvis into which the major calyces (3) open. The minor calyces open into the major calyces. Dark areas called the pyramids can be seen with their apex projecting into a minor calyx. By drawing a line along the outer borders of the pyramids one can divide the kidney into an outer cortex and an inner medulla. The renal artery enters and the renal vein leaves at the hilum on the concave medial border.

In front of the right kidney the second part of the duodenum lies over the hilum and medial border and the suprarenal gland lies on the upper pole. The liver is in front of the upper part of the kidney and the right colic (hepatic) flexure is in front of the lower part. Posteriorly the kidney is crossed by the twelfth rib. Above the twelfth rib it lies on the diaphragm and below the rib it lies on the muscles of the back wall of the abdomen.

Q. By means of a labelled diagram of the urinary tract show (a) where infections may occur, (b) where stones may be found, (c) the possible sources of blood in the urine of an elderly man. (These are three separate questions.) (pp. 155–156, 158–159; Figs.97, 98)
A. (a) Using Fig.98a arrows should point to (1) the kidney itself (nephritis) (2) the beginning of the ureter, that is, its pelvis (pyelitis) (3) the bladder (cystitis). In Fig.97 an arrow should point to the urethra (urethritis) and the prostate in Fig.97a (prostatitis).

(b) Using Fig.98 arrows should point to (1) the kidney itself (2) the junction of the pelvis of the ureter and the ureter itself (3) the ureter as it crosses the brim of the bony pelvis at the sacro-iliac joint (4) the entry of the ureter into the bladder (5) the bladder.

(c) Using Figs.97a, 98a arrows should point to (1) the kidney (2) the bladder (3) the prostate.

Q. Give an account of the anatomical and physiological basis of 'renal colic'. (p. 158; Fig.98a)
A. Renal colic is a very severe pain due to spasmodic contractions of the muscle of the ureter. This is usually due to a urinary calculus

(stone) which is wedged in the ureter most commonly in one of its narrowest parts – at the junction of the pelvis of the ureter and the ureter itself, where the ureter crosses the pelvic brim at the sacro-iliac joint and at the entrance of the ureter to the bladder.

The main severe pain is felt in the loin. Because the same nerves supply both the ureter and also the skin in the inguinal region, scrotum (or labium majus) and inside of the thigh, the pain in the back is often described as shooting down into these regions.

CHAPTER 10 THE GENITAL SYSTEM

Q. What are the structure, anatomical relations and functions of the prostate gland? (p. 162; Figs.97, 101, 103a)

A. The prostate gland lies immediately below the back of the urinary bladder which lies in the front of the pelvis. It is about the size of a walnut and measures 2–3 cm in all directions. The urethra, the channel leading from the bladder to the exterior, passes through the prostate in a downward and slightly forward direction. This part of the urethra is called the prostatic urethra. The anal canal and lower part of the rectum lie immediately behind the prostate.

Entering the upper back part of the prostate are the (common) ejaculatory ducts, one on each side. They pass downwards and forwards and open into the back wall of the prostatic urethra.

The prostate itself is a glandular structure with a stroma of fibromuscular tissue. Its secretion forms the major part of the seminal fluid, but the functions of the constituent parts of the secretion are not well known. The seminal fluid contains important hormones called prostaglandins, hence their name, but these are now thought to come from the seminal vesicles. The secretion also contains a relatively large quantity of acid phosphatase. Its secretion is squeezed out of the prostate during an ejaculation due to the involuntary contraction of its smooth muscle.

The prostatic urethra is the channel for the passage of urine and seminal fluid to the exterior. The prostate frequently enlarges after the age of 60 years and can cause obstruction to the flow of urine.

Q. Describe the structure and position of the uterus. (pp. 167–168; Figs.97b, 104, 107a)

A. The adult uterus is about 7·5 cm long, 5 cm wide and 2·5 cm from front to back. It is described as pear-shaped with the wide end (the fundus) above. The main part is called the body and the lowest part (about one third) projecting into the vagina, is called the cervix. The uterus lies in the pelvis behind the bladder and in front of the rectum.

The uterus consists mainly of smooth muscle but the cervix has, especially at its lower end, a considerable amount of fibrous tissue. The uterus has a cavity which is lined by a glandular epithelium called the endometrium. This undergoes cyclical changes and in the absence of a pregnancy is shed every month between the age of puberty and the menopause. The lining of the cervix is not shed. The uterus is covered with peritoneum. Opening into the upper lateral angle of the uterus on each side is the uterine tube (Mullerian duct). This enables the ovum to pass from the ovary to the uterine cavity. The uterus is usually bent over the upper surface of the bladder. This is called anteversion. This is accentuated by the bending forwards of the body of the uterus on the cervix. This is called anteflexion. In many women the uterus is bent backwards. This is called retroversion and need not be regarded as a pathological condition.

Q. An elderly woman has a prolapse of the uterus. By means of anatomy and physiology explain the signs and symptoms which may be found in this condition. (pp. 167–8; Figs.97b, 104, 107a)
A. During childbirth the fibrous tissue round the cervix of the uterus and the muscles of the floor of the pelvis which keep the uterus in position are stretched and may be torn. This leads to a descent of the uterus and the upper part of the vagina (prolapse). The bladder and urethra lie in front of the uterus and vagina, and the rectum lies behind. The tissues separating these organs from the uterus and vagina are also torn so that the bladder and urethra may bulge into the front of the vagina and the rectum into the back of the vagina.

The patient complains of discomfort in the pelvis and perineum because of the descent of the uterus and the bulging of the bladder etc. Infection of the cervix is common and she will complain of a vaginal discharge.

Because of the tearing of the tissues, control of micturition is affected. The commonest complaint, and the most distressing, is that of stress incontinence. Coughing, due to increase in the intra-abdominal pressure, results in a small jet of urine being expelled because the muscles of the floor of the pelvis are incompetent. Frequency of micturition is common. There may be a complaint of disturbed bowel function.

Q. Give a brief account of the physiology of menstruation. (pp. 164–167; p. 231; Fig.106)
A. Menstruation refers to a part of the menstrual cycle and normally lasts about 4 or 5 days during which time there is a variable amount of vaginal bleeding. The question probably refers to the menstrual cycle which lasts about 28 days but can be as short as 25 and as long as 30 days. The cycle involves changes in the lining of the uterus (the endometrium). If one assumes that the cycle begins with the onset of menstruation then there is vaginal bleeding for about 5 days because the lining of the uterus is shed and the vessels supplying the endometrium are open. After the 5 days' bleeding, the lining is replaced and the columnar epithelium and glands are re-formed during the

next 9 days. This is due to the effect of the oestrogen produced by the ovary especially by the developing ovarian follicle in the ovary.

At about the middle of the cycle, actually 14 days before the onset of the next menstruation, the follicle ruptures and the ovum is set free. The cells of the wall of the follicle proliferate and produce a corpus luteum (yellow body) which in turn produces a hormone, progesterone. This acts on the lining of the uterus which becomes thicker due to a multiplication and engorgement of its cells, enlargement of the glands with an increase in their secretion and an increase in vascularity and fluid. If the ovum is fertilized it develops and embeds itself in the lining of the uterus which is prepared for this. If the ovum is not fertilized the lining is shed at about the 28th day of the cycle and menstruation occurs.

The changes in the ovary are under the control of the pituitary gland and the pituitary gland is under the control of the hypothalamus a part of the forebrain above the pituitary and below the thalamus. The pituitary gland produces follicle stimulating hormone (FSH) which is responsible for the ripening of the ovarian follicle during the first half of the menstrual cycle, and also luteinizing hormone (LH) which is responsible for the development of the corpus luteum.

CHAPTER 11   THE NERVOUS SYSTEM

Q. Describe the external appearance of the brain. (Figs.110a, 119, 124, 130)
A. The brain consists of the forebrain, midbrain and hindbrain. By far the largest part of the brain is the cerebral hemispheres of the forebrain which cover almost the whole of the rest of the brain. The brain is therefore best looked at from below. In the middle posteriorly the upward continuation of the spinal cord is seen as the medulla oblongata of the hindbrain. This ends where the pons, running transversely, is seen passing on each side into a cerebellar hemisphere. The medulla oblongata, pons and cerebellum form the hindbrain of which the cerebellum is the largest part.

The midbrain cannot be seen easily as it passes upwards from the pons into the forebrain.

The cerebral hemispheres, like the cerebellum, consist of outer grey matter, the cerebral cortex, and inner white matter. The cerebral cortex is seen to be very much folded with grooves of varying depth called sulci. The areas between the sulci are called gyri. Some of the gyri are associated with specific functions for example the inner surface of the back of the hemispheres is associated with vision.

CHAPTER 12   THE EYE AND THE EAR

Q. By means of anatomy and physiology describe the condition of cataract. (pp. 221, 223–224; Fig.140a)

A. Cataract is a condition in which the lens of the eyeball is opaque. Normally light passes through the cornea, the clear part in front of the eyeball. It then passes through clear fluid and then through the pupil, the hole in the middle of the iris, the coloured part of the eye. Behind the pupil is a circular biconvex structure called the lens which normally focusses an object onto the retina.

If the lens becomes opaque, as is not uncommon in old people, the light entering the eye is obstructed and vision becomes blurred. Finally all vision is lost. It is possible to remove the opaque lens by an operation and replace the lens of the eye with an appropriate lens in a pair of spectacles. Vision can thus be restored. In recent times there have been attempts to remove the lens of the eye before it is completely opaque and also to replace the lens with a lens placed in the eye.

Cataract can also be congenital or develop as a result of injury to the lens. Treatment of these types of cataract is much more difficult.

Q. With the help of a labelled diagram explain the physiology of vision. (pp. 220–224, Figs.132, 140, 141)

A. Light passes through the cornea which is the transparent part of the outer coat of the eyeball, the sclera. The light then passes through the pupil, the hole in the middle of the visible coloured part of the eye. The light is then focussed by means of the lens on to the retina, the innermost layer of the three coats of the eye.

The light has to pass through the outer parts of the retina in order to stimulate the rods and cones which are the light sensitive structures of the retina. An impulse passes from the rods and cones to the bipolar cells and then to the ganglion cells of the retina. The fibres of the ganglion cells all converge on to one area of the retina, the optic disc, and emerge from behind the eyeball as the optic nerve. The optic nerves pass backwards and meet in the optic chiasma, where the inner halves of the optic nerves cross to the other side. The fibres then go to the visual cortex via a synapse in the back of the thalamus. The visual cortex is on the medial surface of the occipital lobe of the cerebral hemisphere (Fig.125).

The light also has to pass through the aqueous humour (between the cornea and the lens) and the vitreous body (humour) which lies behind the lens. The cones are used in bright light and for colour vision; the rods are used in dim light. The macula lutea (yellow spot) of the retina lateral to the optic disc has only cones. It is used when looking directly at an object.

Q. With the aid of a labelled diagram explain normal hearing. (pp. 225–229, Figs.142, 143)

A. The ear is divided into an external, middle and internal ear and lies entirely in the temporal bone. The external ear consists of an auricle or pinna and the external auditory meatus. The pinna projects more or less backwards from the side of the head. It is usually immobile in human beings. The meatus passes medially towards the tympanic membrane which separates the external ear from the middle ear.

The middle ear is like a matchbox standing on its narrow end. Passing across the middle ear from the tympanic membrane to its medial wall are three ossicles (small bones) called the malleus, incus and stapes. The stapes, like a stirrup, has its footpiece in the medial wall of the middle ear which separates the middle ear from the internal ear.

The internal ear consists of a membranous labyrinth lying within a hollowed out area in the temporal bone called the osseous labyrinth. The osseous labyrinth has a posterior part consisting of semicircular canals, a middle part, the vestibule, and an anterior part, the cochlea. Within the membranous labyrinth is a fluid called endolymph and between the membranous and osseous labyrinths is a fluid called periplymph.

Sounds pass down the external auditory meatus and cause the tympanic membrane to vibrate. These vibrations move the ossicles. The movements of the stapes produce waves of movement in the perilymph. The cochlea contains the cochlear duct. which winds round the inside of the cochlea in a spiral manner. The waves in the perilymph move the basilar membrane of the cochlear duct. On the basilar membrane is a structure called the organ of Corti into which the fibres of the cochlear nerve pass. Movements of the basilar membrane result in stimulation of the cochlear nerve which conveys impulses to the brain. These impulses are conveyed to the auditory cortex in the temporal lobe of the cerebral hemisphere where sounds are heard and recognized.

(Using a labelled diagram it is possible to omit the first three paragraphs of the answer to this question.)

CHAPTER 13   THE ENDOCRINE GLANDS

Q. (*a*) *Describe briefly the thyroid gland and its functions*. (pp. 232–233, Fig.145)

(*b*) *By means of anatomy and physiology explain the clinical features of thyrotoxicosis in a young woman.*

(*c*) *By means of anatomy and physiology describe the possible complications following partial thyroidectomy.*

A. (a) The thyroid gland is the largest of the endocrine (ductless) glands. It consists of two pear-shaped lobes lying in the front of the neck on either side of the lower part of the larynx and upper part of the trachea. Each lobe is about 4 cm long and about 2 cm wide. The lobes are joined in their lower parts by a transverse isthmus, about 1 cm wide, at the level of the upper tracheal rings.

The gland produces two hormones, called thyroxin and tri-iodo-thyronin. Both act on all the cells of the body and stimulate their metabolism so that oxygen is utilized and heat is produced. Our knowledge of the effects of these hormones is based on diseases of the gland in some of which excess hormones are produced and in some too little. In diseases with excess hormones (hyperthyroidism), such as exophthalmic goitre, the patient has a rapid heart, loses weight, is excitable and sweats excessively in circumstances in which normal people would not sweat.

In diseases in which there is too little thyroid hormones (hypo-thyroidism) the patient complains of feeling cold in circumstances in which normal people do not complain of this, puts on weight, has a slow heart and is mentally dull and inactive. This is seen in myxoe-dema and suitable treatment with thyroxin can restore such a patient to an almost normal condition. A cretin is a child born without a thyroid gland and unless treated with thyroxin the child will be stunted mentally and physically.

The determination of the basal metabolic rate (the amount of oxygen used by an individual in a unit of time in quiet, restful, reasonably warm surroundings expressed as the number of calories used per hour per square metre of body surface) is used to diagnose both hyperthyroidism and hypothyroidism. In the former it is raised and in the latter it is lowered.

(b) In thyrotoxicosis there is an excessive production by the thyroid gland of its hormones. Since these hormones stimulate the metabolism of the tissues of the body, it can be said that in this disease the tissues of the body are overactive. Graves described five basic signs and symptoms. These are (1) a rapid heart rate, (2) an enlarged thyroid gland, (3) loss of weight, (4) tremor of the outstretched fingers and (5) exophthalmos – a protrusion of the eyeballs. The last mentioned, if not excessive, is better described as a staring of the eyes. In addition the patient is also easily excited, usually sweating and complains of feeling hot. All of these symptoms can be explained in terms of physiology except for the swelling in the neck, the site of the thyroid gland. This can be explained anatomically.

(c) If the patient has been properly prepared the complication of a 'thyroid crisis' should not occur. In this condition the heart beat becomes very rapid and heart failure may ensue.

Haemorrhage from the site of the operation can be a serious complication because of the pressure on the trachea resulting in respiratory obstruction. If serious, immediate relief is obtained by opening the wound.

Because each thyroid artery is accompanied by a nerve, one or more nerves supplying the larynx may be cut. Most commonly this affects the voice in some way but occasionally irritation of some of the nerves may cause spasm of the vocal folds with respiratory obstruction.

If too much of the thyroid gland is removed the patient may suffer from lack of thyroid secretion and will become myxoedematous.

All four parathyroid glands, embedded in the back of the lobes of the thyroid gland may be removed. In this unlikely event the patient will suffer from tetany manifested by spasmodic contractions of striated muscle.

Q. Give a brief account of the secretion and functions of insulin. (p. 152)
A. Insulin is a hormone produced by the pancreas, a large gland situated on the posterior abdominal wall behind the stomach. The pancreas is therefore mainly in the upper left part of the abdomen. The pancreas is both an exocrine and endocrine gland. As an exocrine gland it produces the pancreatic secretion which is conveyed by a duct into the duodenum to assist digestion. The endocrine secretion passes straight into the blood stream and is produced by groups of cells called the islets of Langerhans situated in the middle of the lobules forming the exocrine part of the gland. The hormone produced by the islets is called insulin (insula means an island).

Insulin is necessary for the proper metabolism of sugar in the body. Sugars (carbohydrates) are absorbed from the small intestine almost entirely in the form of glucose which passes to the liver, where some of it is stored as glycogen. Much of the glucose is also stored in skeletal (striated) muscle as glycogen. This storage of glucose is controlled by insulin. Lack of insulin, which occurs in the disease called diabetes mellitus, results in an excess of glucose in the blood (hyperglycaemia). This results in glucose being excreted in the urine (glycosuria).

Insulin is also involved in the use of glucose by all the cells in the body. Furthermore the proper metabolism of fats depends on the proper metabolism of sugars so that in diabetes mellitus there is a disturbance in the utilization of fats by the body.

CHAPTER 14   THE SKIN

Q. With the help of a labelled diagram describe the structure and functions of the skin. (pp. 7, 235–239; Figs.4b, 6, 147)
A. (This is a question on which students may write too much. For the time allowed the following would be an adequate answer.)

The skin is the outer covering of the body and consists of (a) an outer epithelium called the epidermis, which is a keratinized, stratified, squamous epithelium, and (b) an inner connective tissue structure called the dermis. The epidermis consists of a number of layers the deepest of which is the germinative layer, that is, it produces all the cells of the rest of the layers. As one passes towards the surface the layers of cells produce a substance which eventually becomes keratin, the outermost layer. The cells also die as they approach the surface.

The epidermis also produces the hairs, sweat glands and sebaceous glands of the skin as well as the nails of the digits. It also has cells which can produce pigment.

The dermis contains, besides the usual constituents of connective tissue, a large number of nerve endings and blood vessels.

The functions of the skin can now be related to its structure. The skin prevents excessive fluid loss from the surface of the body and prevents the entry of fluids through the surface of the body. It is protective against a certain amount of pressure and abrasion. The unbroken skin prevents the entry of organisms which are found on the surface of the skin.

The skin is an important sensory organ in relation to touch, pain and changes of temperature. It is important in temperature control both through its sweat glands and blood vessels.

The skin also contains a substance which can be made into vitamin D when the skin is exposed to sunlight.

Q. Give a brief account of the regulation of body temperature. (p. 238–239)
A. The normal activities of the body result in the production of heat which must be lost by the body. Otherwise the body temperature would rise steadily and fairly quickly, that is in a matter of hours. The beating of the heart, the contraction of the respiratory muscles and the activities of glandular structures, especially the liver because of its size, all contribute to the production of heat. Movements of any kind such as walking or even sedentary work result in an increase in heat production.

Heat is lost mainly by the skin. (The way in which this occurs is adequately described on pp. 238 and 239.) This is controlled by the hypothalamus in the forebrain. Part of the hypothalamus responds to the changes in temperature of the blood reaching it and in this way reflexly controls the state of the blood vessels and activity of the sweat glands of the skin.

Some heat is also lost in the faeces and the urine and also on expiration.

Index